BIRDS *in their* HABITATS

For Lou, who shared with me so many of the encounters in these pages, and so much more. Thank you.

Ian Fraser

BIRDS in their HABITATS

Journeys with a Naturalist

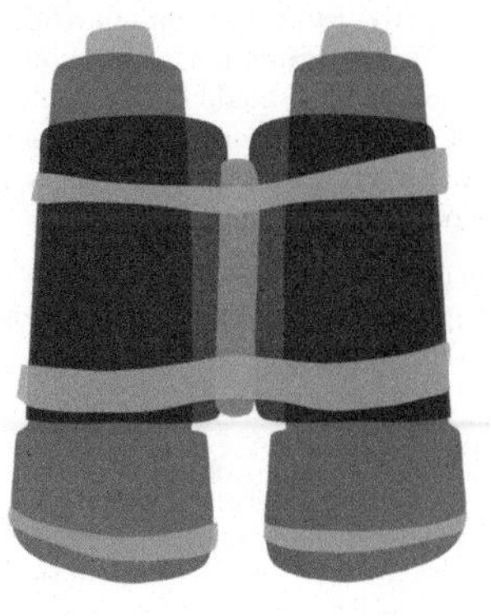

PUBLISHING

A catalogue record for this book is available from the National Library of Australia

Published by

CSIRO Publishing
Locked Bag 10
Clayton South VIC 3169
Australia

Telephone: +61 3 9545 8400
Email: publishing.sales@csiro.au
Website: www.publish.csiro.au

Set in 12/15 Adobe Garamond Pro and Myriad Pro
Edited by Peter Storer
Cover design by Astred Hicks, Design Cherry
Typeset by Desktop Concepts Pty Ltd, Melbourne
Printed in the USA by Integrated Books International

Printed on elemental chlorine free (ECF)
recycled paper containing 30% Post-Consumer Waste
Printed and bound in the USA

Contents

Prelude

The First Time. That's the name of surely one of the most beautiful love songs ever written in English, Ewan McColl's evocative ballad to Peggy Seeger, charting gently and intimately the key stages of a developing relationship. I guess most of us could do that, albeit perhaps more privately and probably less lyrically. But what about our 'other relationship'? Nothing sordid implied there – I'm talking of the love affair that we all share ... with birds. What else can you call the passion that leads us to follow a single bird, not necessarily a 'new' one, for an hour or so to absorb all of its subtle nuances, or just muse in a hide for ages, watching life potter along? Or indeed plan your holidays (or at least attempt to, loved ones permitting) around birds you're hoping to see? Can you trace the tentative beginnings of awareness, through to the full flowering of publicly proclaimed commitment? Was it an epiphany for you, or a gradual and inevitable development? Just pause for a moment there and try to remember.

It seems, when I rescanned their books, that for well-known birders Sean Dooley (*The Big Twitch*), Simon Barnes (*How to be a Bad Birdwatcher; to the Greater Glory of Life*) and Bill Oddie (*Little Black Bird Book*), birds were always part of their lives. For Simon, and perhaps Bill, it was apparently with family encouragement; for Sean it seems to have been despite his father.

For me, I always knew I wanted to do things that involved animals. (I told an infant school teacher that I wanted to be a zookeeper when I grew up; she suggested that a zoo director might be a more appropriate goal. I've no idea still if she was right, but I did become a keeper for a year, though never a director.) My first memories involve following tolerant Sleepy Lizards – which I've learnt to refer to as Shinglebacks since I moved east – through the paddocks north of Adelaide, while Peter the bull terrier kept an eye open for less benign reptiles. He was somewhat neuronally challenged but utterly faithful, and very good at his other job of protecting dad's ducks from stray cats. (My younger sisters were restricted to following caterpillars round the verandah, but, because they too were curious children, it didn't always end well for the caterpillars.)

Being a small boy, what I was really interested in was big African mammals, but by the time I was 12 I was keeping day lists of birds seen on a family holiday. I have the booklet now, all written out in Best, a page per day. It reflects a more relaxed approach to nomenclature – Blossom Parrot, Mountain Duck and Peewee, for instance (though I still use the last two). The only guide, apart from dad, was Cayley's somewhat rudimentary *What Bird is That?*, which might help explain why I recorded Black-breasted Buzzard along the Coorong. The rest of it looks pretty convincing and comprehensive though. Being a thorough boy, I also noted 'Rabbit' and 'Fox (dead)'.

But my epiphany had come shortly before that when we spent a week in a shack at Milang on Lake Alexandrina at the mouth of the Murray River. There was a White-headed Stilt breeding colony just down the road and I was entranced by them, visiting daily to see their impossibly spindly red legs trailing as they whirled crossly into the air, yapping like puppies. To this day, one of my very favourite activities is pottering around wetlands.

It's strange to those of us immersed in the awareness of birds (because after all, it's so much more than just 'watching', isn't it?) that others are – or even could be – unaware of and unmoved by birds. I'm a fan of Simon Barnes' *How to be a Bad Birdwatcher,* in which he says 'I don't go birdwatching. I am birdwatching.' It's not an obscure philosophical credo, but a simple statement of fact that we all know intuitively. Wherever we are (outside at least), there are birds. So whether we're walking to the shops, or hanging out the clothes, or having brunch with friends, or at the footy or cricket or an outdoor concert, we're immediately aware if a raptor or swift whips over, or if a Grey Currawong or Gang-gang Cockatoo calls. It's not something we just go and do sometimes – it's an integral part of our entire existence. Just like being in love really.

So, how was it for you, the first time?

Introduction

One of the wonderful things about birding is all the amazing places it takes us to and I've been pretty lucky in this, especially over the last decade or so. I spend a lot of time wondering while wandering, not only about the birds but about other animals, about plants and, of course, about the places themselves. This book is something of a synthesis of some of those wonderings, in the context of an attempt to share some of the remarkable landscapes and habitats I've found myself in, as well as their inhabitants. I love it when I think I've found some sort of understanding of what I've seen – usually through someone else's insights – but I also love it when I realise that I really can't seem to grasp just what I'm looking at. We are by nature a pretty arrogant mob, and a bit of forced humility does us no harm at all.

The key thing is to acknowledge where we're ignorant – I think the concept of compound ignorance is a significant one. I've seen it described as 'a state in which one not only does not realise his ignorance but considers himself to be knowledgeable', but a friend of mine put it more pithily; 'he doesn't know that he doesn't know'. I might suggest that this is the cause of many of our global problems, but it's probably not my place to do so. However, I'm convinced that it's good to be reminded, for various reasons, how trivial we are in the greater scheme of things, how much there is still to learn about this very wonderful world.

Admitting what we don't know, while simultaneously struggling to make sense of it, is a very important start and perhaps a part of being human – it's certainly a most significant part of birding.

The book is thus also an attempt to share my own understandings, such as they are (mostly, as I say, by reading and trying to assimilate what those wiser and more qualified have to say on the topic), as well as my lack of them. If any part of it sparks a responsive train of thought in you then I've done something worthwhile. If any of the birds and places here bring back memories for you, then that makes me happy. And if by any chance there's anything new in here for you, and you're encouraged by it to go and meander down an unexplored thought track, or go and experience a new place or habitat or bird, then so much the better! Maybe I'll see you out there one day.

Much of what follows reflects my Southern Hemisphere origins and inclinations. I actually think of myself as a Gondwanan, and those lands attract me most. I did spend time some decades ago in Europe, and briefly stopped off in North America on the way home, but since then my wanderings have been almost all south of the equator. If this isn't your part of the world, I really hope you won't be deterred: the principles are universal, many of the fascinating studies I talk about are from the Northern Hemisphere, and anyway isn't it good sometimes to explore somewhere new?

Taxonomy

I have opted to use the International Ornithological Congress (IOC) taxonomy and common names throughout, simply because one needs to be consistent, they are widely adopted, and they provide quarterly online updates, with justifications. I'm using Version 7.2 (as retrieved from http://www.worldbirdnames.org/classification/family-index/ on 18 May 2017).

Acknowledgements

In reality, many people over many years have contributed to the experiences and thoughts herein, not least of them Chris Carter who has sent me to South America many times for work. To a few, however, I owe a more immediate debt, during the actual writing process. If you detect any errors, then they survive despite the best efforts of these people!

- Dr Janet Gardner of the Australian National University read most of the text and tutored me expertly and patiently in the complexities of how birds deal with surviving in a warming world.
- Julian Robinson provided invaluable assistance and advice in preparing my fairly ordinary photos into a state approaching suitability for publishing.

- Associate Professor Naomi Langmore of the Australian National University cast her eye over the material on koels, and told me something that absolutely delighted me!
- Professor Dave Rowell of the Australian National University gave me some excellent tips on the mysteries associated with magpie hybrid zones, thus rescuing me from years of frustration.
- Louise Maher, my partner, read it all as I produced it, casting a very usefully beady eye indeed over the grammar and general presentation. Even more importantly, she was always there for me.
- As I've come to expect, the people from CSIRO Publishing have been encouraging, supportive and helpful from the start. Among them I am especially grateful to John Manger, Briana Melideo, Lauren Webb and Tracey Kudis. Peter Storer, scientific editor, combined tact and ruthlessness to make this a better book; it was a pleasure and an education to work with him.

1
Deserts

Warrigal Waterhole, north-west Queensland: an oasis

Just east of Mount Isa in arid north-west Queensland, a faded sign by the highway indicates a track to Warrigal Waterhole. It is erosion-rutted and at the time we weren't in a four wheel drive, so we left our vehicle to walk the 2.5 km through an ancient bony red landscape of folded hills and plains thickly strewn with gibbers, rounded wind-polished rocks eroded from the hard crust on the hills above. They had been swept from the track to some extent by vehicles, but still posed a rolling threat to the ankles of walkers distracted by a Hooded Robin swooping between bushes. In the distance, the ever-frustrating Crested Bellbird, a key element of the soundtrack of inland Australia, proclaimed eternally 'dee-dee-dee-DEE-dee', constantly turning so that the mellow notes rose and fell, defying us to pinpoint him.

The plains between the hills were punctuated with big red termite mounds and scattered shrubby eucalypts above blonde tussocks of Mitchell Grass and huge spiky hummocks of Porcupine Grass. The termites live by harvesting the grass; it has been said that they are the antelope of Australia and the lizards are the lions and hyaenas (see Photo 1). Arid Australia has a greater diversity of lizards than anywhere else on Earth.

It doesn't sound much like a desert as most people probably interpret it – drifting dunes of burning sand with the only vegetation being date palms around a rare oasis and animals represented by the odd camel train and scuttling scorpion. However, a desert can be any dry place, be it hot or cold; the Antarctic plateau is the world's biggest desert (followed by the Arctic, the Sahara, the Arabian and Gobi

Deserts – so, three of the Big Five among deserts are cold). As a rule of thumb, a desert is defined as any place receiving less than 250 mm of rain a year; more than that, but less than 500 mm, and it's a semi-desert (e.g. USGS 2016). I find the terms arid and semi-arid to be less loaded, but in the end these things are just human conceits. By these criteria, the Mount Isa area, with 460 mm of rain a year, is near the moister end of semi-desert, but a simple statement of annual rainfall isn't the whole story: 85% of The Isa's rainfall comes in the Southern Hemisphere summer months of November to March, when the average daytime temperature is 36°C, so most of the rainfall evaporates before becoming available.

In such areas, the low stony ranges provide essential oases, especially where gorges form relatively cool refuges. In these gorges, as at our goal on this particular walk, water from summer storms may form deep pools in narrow slots where the sun doesn't penetrate and evaporation rates are low. These waterholes are wildlife magnets – and, as an inevitable spin-off, they are also irresistible magnets to wildlife lovers such as ourselves. This was June and the temperature was barely into the 30s, but we were pleased enough to leave the plain and climb the shoulder of the hill behind which lies the shaded waterhole. Here we were reminded, however, that wildlife in such a place can take many forms. As we descended into the narrow forest of River Red Gums and were about to enter the gorge, we met a large soot-black bull coming out, water still dripping from his lips. In such a situation, he could have been either a feral animal or station stock but, even if the latter, 'domestic' is not a term I'd use for him. His resemblance, with wicked black-tipped horns, to pictures I've seen of Spanish fighting bulls added to our unease; we climbed back up the stony hill among the Porcupine Grass, but fortunately he was in a benign mood and ambled off into the trees. Birding isn't always the genteel pastime that you might suppose!

We settled onto a shelf of granite rock above a sandy beach and waited for the birds to come. I suppose such birding must be a bit like playing the poker machines – you can't know what's going to happen next, but in the case of the birding you absolutely can't lose. Even if nothing had come, we would have a peaceful hour absorbing the still beauty of a special place.

As it turned out, this location wasn't still for more than a few seconds. There are very few places in all of mainland Australia where you can't find a Willie Wagtail: a bold, nosy and opinionated black and white member of the fantail Family. (The Family, found from India to Australia and New Zealand, was once thought to be related to the Northern Hemisphere flycatchers, but we now recognise them as part of the great assemblage of songbirds which evolved in isolation in our part of the world.) This Willie was a youngster, as attested by its buffy eyebrows; later they will turn white, but this one wasn't too young to investigate us closely and pass chattering comment in the way of Willie kind. However, Willie wasn't attracted by the water for its own sake, but by the insects the water also drew. Meat eaters, be their prey insects or vertebrates, don't often need to drink (though they may sometimes choose to do so), because there is enough metabolic water in their food: this is an excellent adaptation to living in the dry lands.

Desert seed eaters

The desert birds that do need to drink daily are those that eat seeds, which have very little water content. In Australia, this overwhelmingly means finches, pigeons and doves, and parrots. And at Warrigal Waterhole it wasn't long before a familiar tootling chorus, like lots of toy trumpets, brought a shared smile as we anticipated the coming of one of our favourite birds, the delightful Zebra Finches, ubiquitous across inland Australia. In central Australia, the Pitjantjatjara people hear their cheery nasality in much the same terms that we do – to them these finches are *Nyii Nyii*.

As they streamed into the bushes above Warrigal Waterhole before coming down, we took in anew their characteristic orange-red bills and legs, white faces bordered by a black streak running down from the eye like errant mascara, and broadly striped black and white tails; in addition, males have barred breasts, rusty cheeks and white-spotted rusty flanks. They are delightful.

They are also astonishing little calculators, constantly assessing their environment and determining where to roost to be in reach of the morning's breakfast bar, when it's time to move on, and when to start breeding. Day-to-day living is tricky enough: as well as needing to drink daily, a 'zebbie' needs to eat up to 6000 tiny grass seeds each day, though fewer if bigger ones are available (Zann and Straw 1984). Both water and seeds are not necessarily available in the same place, however; although Warrigal Waterhole will probably always be here in all but the most vicious of droughts, the seeds represent a much more ephemeral resource. If the little finch can't find both of these things on every single day, it will surely die. So, it must be constantly planning, not just where to roost tonight to be in reach of water and seeds tomorrow, but where it will go for a seamless transition when the pool dries up or the seeds finish.

Breeding is even trickier, made more so by their choice of baby food. Like many vegetarian birds, some other grass-finches feed their chicks on insects to give them a high-protein kick-start to life, but others, including zebbies, are purists – they start their offspring off as they'll continue, on seeds. The only concession they make is to provide soft unripe seeds, which aren't nearly as easy to find as ripe dry ones (Payne 2017; Payne and Bonan 2017). Seeds are not like insects, which appear almost as soon as the rains do, so zebbies planning to breed have to make some pretty complex calculations. If the weather is cool, grasses take longer to grow and set seed than if it's hot. The soil moisture level must also be taken into account; this is a function of both the time since the previous rains and what the temperatures were in that time. The end result is that the tiny birds must remember how long it is since the rains came and, as a result of what can only be described as complex computations, start to breed anywhere between 4 and 12 weeks later. We have only recently started to unravel this amazing narrative (and for more on zebbie ecology, truly one of the great Australian stories, I'd strongly recommend Morton 2009).

Other grass-finches joined the stream of zebbies at the water's edge in the next hour too, among them a few very smart Painted Finches (red and brown above with white-spotted black sides), a couple of owlish-faced Double-barreds (with two black bands across a white underside, blue bills and spotty wings), and some unexpected Long-tailed Finches (red bills, grey heads, black throats and long pointy black tails). Here, the Long-taileds are at the very eastern edge of their range. The grass-finches (Family Estrildidae) apparently arose on the grassy African plains and, like humans much later, expanded east through

Asia (or possibly they began in India and spread both west and south). As Australia approached Asia closely enough in the past few million years, new passengers hopped aboard, including rodents, bats, modern crocodiles, swallows – and grass-finches. It seems that they arrived in three waves, the first being the firetails, including the Painted Finch. Next came the ancestors of zebbies and Double-barreds, and most recently another group, including the Long-taileds, most of which haven't left the tropics. (There is also a Timor Zebra Finch, which may have derived from some perhaps dissatisfied Australian settlers that later made the reverse crossing, or from some that chose to stay there rather than continue south.)

A bit of a rule of thumb in Australia is that within a group of animals the oldest members are those that have penetrated the deserts, but this doesn't really hold true with the grass-finches. Of the firetails, only the Painted Finch lives in the dry country, but it survives solely in the sheltering ranges. Of all the Australian finches, only the zebbie has truly penetrated the deserts. (Of course, it could be that this model of the order of arrival in Australia is incorrect too.)

Meanwhile, back at the waterhole, we were watching the zebbies drink, along with a couple of Painted Finches that had slipped in along the gorge wall. Their approach to drinking is quite different though: the zebbies stick their bill into the water and suck, which is most unusual among birds, while the Painteds more conventionally take a bill-full of water and tip it back, over and over again. The sipping zebbies can get their fill much more quickly than the tipping Painteds, and consequently need to spend less time at the hazardous water edge (e.g. Forshaw *et al.* 2012). While we were at the waterhole, it's likely that our presence was deterring predators, but Australian Hobbies, Peregrines and Black Falcons, Collared Sparrowhawks and Brown Goshawks all hang around such watering points waiting to transform an unwary bird into a meal. It seems that a few, mostly tropical, Australian grass-finches are the only ones of their Family to have developed this useful sucking skill, despite the fact that many relatives also live in arid lands in Africa and Asia.

Better known suckers of water are the pigeons and doves, and by now two desert doves were arriving in small groups and creeping nervously to the water. Peaceful Doves and Diamond Doves are found throughout most of the inland, though as seed eaters they too need to drink daily. Both are small and elegant: the Peaceful has a grey barred body and a blue eye-ring; the Diamond has white-spotted brown wings and a red eye-ring. The high-pitched 'toodle-oo' of the Peaceful Dove is a soporific sound on a warm afternoon by any creek line.

Below us little groups of doves dipped their bills in and filled up quickly, whirring off as soon as they finished. Superficially it seems that they're doing just as the zebbies are, but in fact their approach is quite different again. The finches use their tongues as a double-action scoop, at up to 20 times a second, taking a droplet of water into the mouth, and from there back into the oesophagus and crop, via the pharynx. (The crop is a very useful sack, an extendable side-wall to the oesophagus for carrying seed away to be digested later at leisure and in safety.) The doves, however, create a sort of peristaltic pump by sending waves of muscular contractions along the oesophagus to pull the water back (e.g. Elphick 2014). There is nearly always more than one solution to any evolutionary question and, with enough time, nature will come up with different answers in different situations.

Eventually, as we had hoped, a flock of another of our favourite birds (of which there are, admittedly, rather a lot) suddenly swirled into the branches above us, like leaves sprouting from the sere twigs. It saddens me how many Australians don't realise that Budgerigars are not only wild birds or parrots, but even Australian, and can be seen in huge flocks pretty much anywhere in the inland where conditions are good. It saddens me even more to see a budgie in a cage – for a bird that is essentially sociable and flies constantly across the vast plains, solitary confinement must be a harsh sentence.

No organism, be it bird or farmer, can survive in Australia without adapting to the integral unpredictability of our climate, driven by the El Niño cycle (or, more formally, El Niño–Southern Oscillation, or ENSO to its friends). Budgies are true children of El Niño.

Budgerigars

The budgies at Warrigal Waterhole dithered and panicked and eventually mostly moved on, with only a few coming to drink; maybe there was after all a bird of prey loitering with intent, or maybe it was just us. We left soon afterwards and I'm sure they came back.

El Niño

In the first decade of the 20th century a young Australian called Dorothea Mackellar who was travelling in Britain and Europe published a poem called 'My Country'. It is in the form of one side of a conversation with someone who loves the misty European countrysides, acknowledging that love while expressing her own passion for the harsh extremes of Australia. She was from a wealthy Sydney family, but spent formative time on a family property in the mid-west of New South Wales. The poem caught the Australian imagination and was widely printed in newspapers. It has been put to music more than once – most recently, and somewhat mind-bogglingly, by eminent Australian composer Elena Kats-Chernin for the Vienna Boys Choir! I recall singing a version at primary school and the tune is lodged in my brain, though I can't determine what its origin was. To many of us this is still the unofficial national anthem. In it she characterises Australia as a land of 'drought and flooding rain'. And there she nailed it: a truth that some people, even among those who manage the country to a greater or lesser degree, still don't get.

When, every few years, the winds fail that normally blow west across the Pacific and keep warm ocean water 'mounded up' around Australia, El Niño is triggered. A layer of warm water 100 m thick flows 'downhill' to South America where it prevents the normal rich cold upwelling of nutrients by the Humboldt Current on the west coast of Peru. The surface algae that need these nutrients die, and the abundant tiny anchoveta fish starve. Normally, they support one of the world's great fisheries; population crashes of seabirds and marine mammals follow. On the desert coast, storms and floods scourge the land. Meantime across the Pacific the opposite is happening. The waters off eastern Australia cool and the rains fail. Droughts follow and may affect the entire continent; ferocious fires almost inevitably follow. If there is a series of ENSO events, drought can last for a decade, as happened in the 1980s and for the first decade of this century. The final weakening of the system, which generally happens after about a year, follows from the re-establishment of the easterly winds; it is often followed by torrential rain and flooding in inland, as well as coastal, Australia. This is the La Niña part of the cycle. (In South America, La Niña, as we would expect, brings drought.) Of course, other parts of the world are affected too (and other engines drive Northern Hemisphere weather and Indian Ocean systems) but only Australia is affected continent-wide by ENSO.

This is very much a simplification of a very complex process that we are still yet to fully understand – we only began to do so in the 1980s – but this is the essence of it. In south-eastern and south-western Australia, European concepts of season work some of the time – until ENSO rolls in. In the Australian arid lands, such concepts are never relevant. It may not rain at all, or not meaningfully, for years, then the land may be a vast shallow lake. As the water soaks in, some plants and animals may reproduce for the first time in years. No matter the time of year, you have to be ready to go when the rains finally come. One way of dealing with ENSO is to build populations up when conditions are right, and trust that at least some of the hordes of scions make it through the coming drought to breed again when the rains return.

The breeding of wild budgies has to be seen to be believed. Along the normally dry watercourses, every hollow in every River Red Gum has a nest; the sky is a swirling shroud of green and the ground is covered with tiny green and yellow parrots garnering seeds. Just one storm during a drought can trigger such a frenzy of activity. If conditions are right, a budgie can begin breeding when just 4 weeks old, so that a 6-week-old bird is already hatching her own eggs (Forshaw 1994). Most birds defer beginning incubation until the last egg is laid (at one per day); until that happens, an egg can stay in 'suspended animation' without coming to harm. A budgie begins to incubate immediately she starts to lay (up to eight eggs on alternate days), so nestling development may be staggered by nearly a fortnight (van Dyck 1995), the younger birds being given a more liquid regurgitated diet. This is so that, if the bonanza fades rapidly, at least the early hatchlings might get through. In addition, if the good times are still rolling on, she can start laying again after one clutch hatches (see Photo 2).

As a result, numbers build up to enormous levels in good conditions, and they die by the millions in severe drought. One central Australian landholder in 1931 removed and burnt almost 5 tonnes of dead budgies from one dam during a heatwave (Chisholm 1958), but the safety net of vast numbers means that at least some hardy, canny or just lucky budgies are around to take part in the next breeding frenzy. *This* is what 'droughts and flooding rains' or 'boom and bust' is all about.

As we walked out from Warrigal Waterhole across the plain, the sun was much higher and so was the temperature, but that was forgotten as out of the Porcupine Grass (widely known as Spinifex) materialised a glorious Spinifex Pigeon. They always seem to perform this trick of emerging from a red landscape by some sleight of feather: a movement changes your focus and suddenly one, two, a dozen exquisite rusty little birds have been there all along. Their ludicrously tall head tufts and red, black and white visages that seem to have come from a child's face-painting stall are borne swiftly through the stony landscape on legs that give the impression more of whirring wheels. Some birds really test one's resolve not to be anthropomorphic.

Near to the vehicle, the Crested Bellbird, which we had failed so dismally to locate despite its constant calling, had apparently tired of the game, and bounced energetically across the ground chasing a fluttering moth alongside the track. It's a funny business, this birding.

South-west Queensland: locust swarms

Another year, and further south in Queensland, in the channel country, we were driving through vast plains braided with numerous (mostly dry) stream lines merging and splitting. We had been watching the great swirling clouds approaching, then suddenly drove into one, the windscreen abruptly spattered with yellow waxy material sprouting legs

Locusts

This is a remarkable story in itself. 'Locust' is a curious term referring not to a species, but to something much harder to define: some grasshoppers can be locusts sometimes. In Australia, just four arid land grasshopper species can perform the trick, but there are others elsewhere in the world. They are normally solitary and sedentary, but in some specialised conditions they suddenly multiply dramatically and start to move in vast numbers, eating hundreds of tonnes of vegetation per day. Individuals change in colour, size and behaviour when in high densities compared with when they are solitary, which is known as kentromorphism.

It is all to do with the uncertainties of living in deserts, especially those ruled by El Niño. Normally the female lays ~40 eggs in a hole drilled into open ground between patches of vegetation, scattered over vast areas. In drought, the ground can be too hard even for them, and they are forced into smaller and smaller moist areas such as at the foot of dunes or in soaks. Eggs are laid closer and closer together, until there may be 3000 holes per square metre (i.e. there may be a thousand million eggs per hectare – that's a lot of little grasshoppers!). What's more, they only hatch in favourable conditions (no point in emerging in a drought), so they may sit in the ground for years until the right conditions come along, then all hatch simultaneously. When such large numbers hatch at once, a behavioural change occurs, apparently induced by developing in close proximity to lots of other little grasshoppers. In such circumstances, they seem to need company, which they don't if they have hatched separately. They also look different – darker pigments form in the skin, so they absorb more of the sun's heat, so are more active, so in turn must eat more. All this leads to the formation of a swarm and forces them to keep moving, so they must keep eating. Under normal circumstances, these great swarms are driven by the winds into south-eastern Australia; when the wind speed exceeds 11 kph, they move with it. As they move into cooler areas, they stop moving and eventually all die. Meantime they have been laying eggs as they go, which, until the next drought changes things, will hatch as 'normal' grasshoppers (e.g. Symmons 1985; Rentz 1991).

and heads. The Spur-throated Locusts were flying, so I slowed to a speed at which their rocky bodies just bounced off the glass.

Of course, such a bonanza of protein attracts hunters in vast numbers too. The locust cloud carried within it hundreds, probably thousands, of White-browed and Masked Woodswallows, gorging until they could eat no more. But if you take thousands of locusts away from a swarm of millions, you're still left with millions ... A couple of Brolgas darting to and fro on the plain were similarly overwhelmed. Above them a flock of some two dozen Brown Falcons circled and snatched. I think of these birds as being typically in pairs, and this was something I'd never seen before.

Woodswallows and nomadism

These two woodswallow species are as much children of El Niño as are the budgies and zebbies. There are six Australian species – one of these extends into the Pacific and South-East Asia. Despite the name, they have no relationship with swallows, but do share with them the characteristic of catching insects in the air; their closest relatives are the Australian magpies, butcherbirds and currawongs. This has only recently been accepted though: in 1973, JD McDonald, who had headed up the Bird Department of the British Natural History Museum, wrote in his *Birds of Australia* that 'their connections are obscure' (McDonald 1973). One source of confusion is their honeyeater-like brushy-tipped tongue used for extracting nectar from flowers. Back in 1907, however, a connection between woodswallows and Australian magpies had been suggested by William Pycraft of the British Museum; this was largely ignored, but eventually confirmed and extended to currawongs and butcherbirds by Allan McEvey, from the Museum of Victoria, in 1976. McEvey based his conclusions on skeletal studies (as had Pycraft 69 years previously). In particular, he noted, uniquely among Australian passerines, a shared strange extension of a cheek bone – a 'bifurcated zygomatic process' if you please (McEvey 1976).

All but one of the six Australian woodswallows are found in the desert lands, but only White-browed and Masked are inveterate nomads, following the rains, the grasshoppers or other resources wherever they may lead them.

Migration and nomadism

True migration – the predictable annual movement of all or most of a population from a breeding ground to a warmer wintering ground – is not nearly as significant in Australia as elsewhere in the world. Yet again, the reason is El Niño: there is not much point flying 10 000 km from the far Northern Hemisphere for a warm relaxing break from breeding and finding yourself in a drought. So, although hundreds of millions of birds yearly fly between northern Europe/west Asia and Africa, and between North and South America, the birds that breed in the endless northern forests and tundra of eastern Asia (due north of Australia) also fly south, but stop short of coming to Australia. The exceptions prove the rule – waders and seabirds whose habitats are effectively drought-proofed. (In the winter-cold south-east of the country, many species are true migrants but, having bred in the south, they only go as far as Queensland, or in a few cases to New Guinea and adjacent Indonesia.)

A few other species undertake annual migrations that, quite frankly, seem to make no sense at all (though of course they must). A small number of tropical species – Red-bellied Pittas, Torresian Imperial Pigeons and Buff-breasted Paradise-Kingfishers, for instance – breed in northern Australia in the wet, and pop across Torres Strait to New Guinea for the dry season. Why? Conditions are near-identical. Similarly, Double-banded Plovers breed on the south island of New Zealand in summer, then make the relatively short flight across the Tasman Sea to spend the rest of the year at similar latitudes in south-eastern Australia. Again the advantages of the strategy are not obvious.

Nomadism on the other hand – incessant wandering following rains, seeding or flowering, or just seeking them out – is a very significant part of the lifestyle of many desert species. For inland waterbirds, there is no other way they could survive.

White-browed and Masked Woodswallows may turn up after rain in their thousands in an area they have not been seen in for years, then disappear again. Curiously, they are nearly always together, but they're an odd couple. They have very different plumages, but to our ears their calls are indistinguishable: you might reasonably think this is just us, but more startlingly their mitochondrial DNA seems identical too. This is not supposed to happen – much more often the secrets of the cell reveal two separate species where our eyes had only recognised one. They not only look very different to us but, more relevantly, to each other as, while they regularly breed in mixed colonies, hybridisation is very rare (Joseph *et al.* 2006; Joseph 2009). What does this tell us? No-one seems to be sure actually (other than that they separated from a common ancestor in recent times), which I often find quietly satisfying; a bit of a reminder never goes astray about our need for more humility than we are wont to display.

By now the great black bruised storm clouds to the north, glowing inside with near-constant lightning, suddenly burst upon us, nearly clearing the windscreen of locust detritus with hammering rain: time to drive on.

Flock Bronzewings and Fairy Martins

The previous day, we had been delighted to encounter another nomad a little to the east, quite unexpectedly, with half a dozen somewhat theatrical black and white faces emerging from the saltbush around a little slot dam almost on the outskirts of Windorah on Coopers Creek. 'Theatrical' is actually not such a bad descriptor: the Flock Bronzewing's species name is *Phaps histrionica*, with *histrionica* being from the Latin for a pantomime performer, from its apparently made-up face. I'd never seen Flock Bronzewings so far east, but wasn't especially surprised to do so now. These too are true children of El Niño, appearing in vast numbers in good years and spreading far across the land, studding the ground with their eggs (tales abound of sheep being stained yellow by them, after a night's rest on the plains) then disappearing into some secret fastness. Flocks of up to a million were reported from the 19th century, but by the end of that century it was feared that they may have succumbed entirely under relentless habitat alteration, especially over-grazing by sheep and rabbits, shooting and cat and fox predation on nesting birds and eggs. By mid-20th century, they were making a recovery – perhaps related to rabbit control by the *myxoma* virus (i.e. myxomatosis)? – and I've driven through big scattered flocks in South Australia, but flocks of hundreds of thousands seem but a wistful dream now.

Another bird found throughout the channel country seems to have done much better from human intervention. The roads of the channel country can only exist because of numerous bridges and even more numerous culverts, which might pass beneath the tyres without being noticed were it not for the whirling clouds of birds above each one.

Stop and open the windows and the air is filled with the musical buzzing as even more Fairy Martins arise from beneath us. Their ancestors relied on cliffs, overhanging banks and big tree hollows to anchor their jumbled apartment cities of funnel-mouthed, enclosed mud nests. Today's rusty-headed 'fairies' have a near limitless range of choices: under these bridges and culverts, and under the eaves of a million bush dwellings and sheds. The Fairy Martin is a relative newcomer to Australia – its closest relatives are found in Africa, Asia and the Americas – and it seems likely that only human assistance has enabled it to penetrate the desert lands. A relative, however, the beautiful White-backed Swallow, has its closest relatives in Africa, but has been here long enough to have truly adapted to desert living and to have evolved in different directions so that it is now recognised as comprising its own genus. It has no need of human constructions to nest, but digs its own burrows (or uses those of bilbies or rat-kangaroos) and makes a nest chamber at the end.

The Sahel: arid woodland

The Sahel is a vast swathe of arid woodland separating the Saharan sands from the grassy savannahs further south. Up to 1000 km wide, it crosses Africa from the Atlantic to the Indian Ocean, forming part of a dozen countries. I encountered it on a dedicated bird trip with expert guides in far northern Cameroon, from the northern regional capital of Maroua to Waza National Park. This is a distance of just 120 km, but on crumbling roads (a legacy of long-gone French and British colonial governments and now best thought of as craters connected by remnants of bitumen), and of course bird stops, that can take hours. The few motorbikes carry up to four people – and not a helmet to be seen. Other road users include bicycles carrying almost anything imaginable, including live goats, vast loads of firewood and long bundles of reeds carried crossways, which present serious hazards. Women walk, carrying equally huge loads of wood or big earthenware pots on their heads. The level of firewood extraction is alarming, made more so by the piles by the road in villages, tied with grass for collection by semi-trailers for use in towns. In such a dry climate it can't be sustainable, spelling a grim future alike for wildlife and the people relying on the wood for cooking. It is already noticeable how much denser the thorny acacia scrubland is away from villages.

Cameroon itself is on the cusp between west and central Africa, perched on the 'armpit' of Africa where the west coast suddenly stops heading north from South Africa and swings due west. Vibrant, struggling with official corruption and a president determined to hold onto power (33 years and counting as I write, but at least he was elected), crumbling infrastructure, pretty basic lifestyles, especially in the north, and environmental damage, but a country of great natural richness where people are welcoming of outsiders and Christians and Muslims seem to rub along pretty comfortably.

When we stopped outside a village to search a big volcanic outcrop for birds, we attracted an audience – people watching the watchers. Cyclists stopped to exchange comments on us, and a gaggle of children appeared, along with a patriarchal chap in robes and spectacles, on a very little motorbike. Among the rocks, a handsome stolid Greyish Eagle-Owl, an arid land specialist, glowered at us from within a rock shelter. Further up the road, we stopped to search for a real 'special' though – a Quail-plover – in an unpromising-looking dry sparse shrubby landscape with a permanent haze of dust or smoke, which belied the already fierce heat of the day.

The Quail-plover: of birding and twitching

The Quail-plover is neither quail nor plover (Australia has a particularly bad habit of using such misleadingly chimeric names, but we're not alone in it!). The buttonquails are a Family of birds not at all related to quails (yes, I know, but life is easier if you just accept such things), being more closely related to shorebirds, though they are certainly superficially quail-like, being small and dumpy and prone to exploding out from under your feet and disappearing again. The Family comprises 16 species of the genus *Turnix*, spread across much of Africa, southern Spain, south-eastern Asia and Australia, plus the Quail-plover: a single-species genus pretty much limited to the hot dry Sahel grasslands.

The search was an interesting lesson in different approaches to birding; about half of my companions on this trip, none of whom I knew, were hard-core twitchers, men (all nine of us carried a Y-chromosome) with an already impressive list of birds from around the world and whose primary purpose in life was to add to it. Their competitiveness was mostly subtle, though not always so, and pressure on our excellent guides was always simmering. The Quail-plover was one reason for them to have come to somewhere as remote, and frankly often difficult, as Cameroon, and without seeing the bird we weren't going to leave this harsh plain until dark if necessary – in which case we would have been back the next morning.

I don't regard myself as primarily a twitcher, though I make efforts to see and enjoy new birds, and keep careful lists from my travels, which I enter on databases. However, I never tire of the familiar birds either: the day I get bored with Galahs (an abundant but gorgeous Australian cockatoo) it will be time to hang up my binoculars. I couldn't conceive of blinkering myself to the rest of nature, let alone only to birds I hadn't previously seen. At one stage on the trip, our guide was showing me some lovely epiphytic orchids, and I heard one of our party mutter in his broad northern English accent 'bad as effin' Nature Trek'; it was only partly light-hearted. Were I involved with that tour company I'd have taken it as a compliment, but he didn't intend it as such.

For me, everything about Cameroon was new and interesting. I was revelling in that and, although I wanted to see the Quail-plover too, and joined the line sweeping the plain systematically, I made sure I was on the end. From here I could discreetly skive off from time to time to pursue such Sahel delights as Rufous-tailed and Black Scrub Robins, Chestnut-bellied Starlings, Chestnut-backed Sparrow-larks and smartly coiffured Black-headed Lapwings. Ponderously graceful White-bellied Bustards flapped off and gloriously elegant African Swallow-tailed Kites drifted weightlessly overhead.

In the event, we did find our bird when it flushed from the ground, flying a short distance with its strange bouncy flight totally unlike the rocketing whirr of its relations. We watched it for some 45 minutes while it stayed quietly and sensibly in the shade – in direct contrast to us. When it moved on, I was glad that our guide declined to let us follow it again, so that it could get on with its life.

At another stop, even sparser and hotter than the Quail-plover's, we tried, this time in vain, for another Sahel special: the apparently rare and certainly little-known Golden Nightjar. Later, some of the party would make a long return drive in the evening to try again at the end of a protracted and exhausting day. I couldn't have done it by then, and thought they deserved better success than they had.

Waza National Park and its panting birds

The drive up that crumbling highway was a bit of a trial (the French were building a lovely new one alongside it, which I assume is now open), but my sadness at remembering that trip now has nothing to do with the minor discomfort of the time, which has been entirely overshadowed by subsequent events. At the time, Boko Haram existed in north-eastern Nigeria, but had been quiet for a while and not generally regarded as a threat elsewhere. Then, in February 2013, a group of these insurgents crossed the nearby Cameroonian border by Waza, our destination that day, and kidnapped a French family, releasing them 2 months later amid strongly denied claims that money had been paid to the Boko Haram cause. At that moment, the trickle of visitors to the north, and especially to Waza National Park, stopped dead, and with it the small but valuable income that we were bringing the local economy. Most chillingly, in January 2015, a local bus was stopped on the stretch of highway we drove that day, and at least 11 people were murdered. It seems that their crime was to live in peace with their neighbours. I wonder what has happened to the people we passed on the roadside, the people we waved to, smiled at, bought supplies from. It will be a long time before birders pass that way again.

Waza National Park comprises 170 000 ha of the Sahel: dry and dusty when we were there, with viciously spiny deciduous acacias above sparse dry grasses. And it was hot. Our accommodation comprised a group of basic, but comfortable, round stone cottages clustered on a stony hill at the edge of the park. We got there in the afternoon, and the heat lay on the land like a blanket. The only animals that seemed happy with the conditions were the agamid lizards that scurried about in the open-sided restaurant.

Ethiopian Swallows sat on twigs in the full sun, under which their wing and back feathers glinted blue-black, beaks open, panting (see Photo 3). Above them a glorious White-throated Bee-eater sat doing the same. I love bee-eaters, and this one is typically splendid, with its white face between black cap and bib divided by a broad black stripe back through its eye, ochre back grading to green-blue wings, and an unusual pale aqua breast. The elegantly long tail streamers told us it was breeding. Indeed, this species flies from the great tropical rainforests south of here to breed in the Sahel, relying on the ephemeral winter rains. This seems a risky strategy but, although it is equally arid, this country has a more predictable climate than El Niño-taunted Australia. Why these birds should choose to sit out in the blazing sun is a bit of a mystery, but perhaps there was a subtle breeze – if so, it was certainly too subtle for me to detect! The clustering little tawny African Silverbills (grass-finches, or waxbills, like the finches at Warrigal Waterhole) were sensibly in the shade on branches of a big tree, alternately preening each other and just sitting – and panting.

All this panting was a direct response to the heat. The original purpose of feathers, in non-flying dinosaurs, was insulation. (Later they adapted them to flight, but that is a story for another day.) Preventing loss of heat is still one of their major purposes, and by trapping a layer of insulating air they do the job very efficiently indeed. Birds don't have sweat glands (which we use to enable evaporative cooling) – there would be little point having them under the feathers. Ours work because we have dispensed with insulating body hair. A little water can escape through the bird's skin, but this isn't very efficient. Perhaps the major way that birds prevent themselves from overheating at temperatures much over 30°C is by evaporative water loss through exposing the moist mouth and throat tissues to the air and breathing rapidly to create an air flow over them. Some birds step up the process by 'gular fluttering', rapidly vibrating the lower mouth and throat surface, including that of the upper digestive tract: pelicans and cormorants demonstrate this well. The fluttering is pushed by the bony hyoid, which supports both the tongue and the upper respiratory and digestive tubes. Because this process affects the digestive tract, rather than the respiratory one, the dangers of hyperventilation are reduced. Gular fluttering seems to be especially efficient in bigger birds. What is odd is that it doesn't occur in passerines, which are the largest and most recently evolved Order of birds. It is unlikely that they had the ability and later lost it, so presumably the ancestral passerine hadn't learnt the trick and it didn't ever evolve in this Order.

The swallows and bee-eater also held their wings partly extended to expose the large blood vessels under them to the air, further enabling heat loss. In part, the swallows looked so glossy because they had pressed their plumage close to the body, squeezing out as much of the insulating air as possible. I do wonder why they didn't use the shade though. I guess you need to be tough to live in the Sahel.

I have an abiding memory of a dramatic sunset that night: a huge sun setting white, rather than red, through a bare thorn tree, with a silent flight of Black Crowned Cranes drifting over in silhouette.

Waza: guineafowl and sandgrouse

Next morning we drove into the park, past herds of antelope, big Roans and Korrigum (formerly regarded as a subspecies of Topi) in a landscape seemingly too sparsely grassed to support such a large biomass of animals. A pair of Common Ostriches and a huge Arabian Bustard stalked away from the vehicle.

We stopped near a muddy waterhole and watched as mixed flocks of doves streamed into the nearby thorn bushes, dropped to the ground and walked cautiously to the water's edge. A mob of Helmeted Guineafowl came *en masse*, while keeping a wary eye on us across the water. Guineafowl have always appealed to me: I would be tempted to keep some if their extraordinarily raucous cackling wasn't guaranteed to test the patience of neighbours beyond reasonable limits. Plump round tailless bodies, dark grey spangled all over with white stars, are carried on strong grey legs. Their white face (at least in this west African race) is set off with red bill and wattles, and a funny little cassowary-like casque on the head. This flock isn't just a casual gathering: Helmeted Guineafowl are known to stay with the same flock, which can contain up to 200 birds, for years. Unlike other guineafowl species

Staying cool in a hot land: a tricky balance

As with most problems, there is no perfect solution to the crucial dilemma of staying cool when the surrounds are very hot. Panting works quite well, but, as temperatures rise, so do the spectres of two problems involved with it. If it's dry – as in Waza – the loss of body water is increasingly dangerous. Albright *et al.* (2017) looked at the impact of a 4°C temperature rise on five species of southwestern US desert passerines, with regard to evaporative water loss (i.e. through panting). They showed that small birds lose water at a relatively higher rate than larger ones, so are at higher risk. The basis to this lies in physics. A smaller bird has a relatively larger surface area than a large one – more on this on page 160. But, given that they don't lose water through sweating it seems counter-intuitive that this could be relevant. However, it is not just the body surface that is relatively larger: the area of surfaces such as the upper digestive tract and interior of the mouth from which water is being evaporated is also larger.

On the other hand, if humidity is high, panting simply doesn't work – you can't evaporate water into a saturated atmosphere. However, we're becoming aware of the complexities of another use for the bill in this situation. Bills can dump excess body heat by radiation and convection without losing water, and this works just as well at conditions of high humidity (it seems that this is the major purpose of the ridiculously large bills of the tropical toucans – Tatersall *et al.* 2009). Allen's Rule predicts that such organs used in heat exchange will be disproportionately larger in warmer climates, but only up to a point … As the surrounding temperature approaches the bird's body temperature, radiation and convection cease to work. Even worse, if the surrounds are hotter than the bird, a large bill actually becomes a hazard by *absorbing* heat.

Janet Gardner of the Australian National University and colleagues set out to test Allen's Rule for a wide range of Australian passerines, while also exploring the role of humidity (Gardner *et al.* 2016). They measured 2864 museum specimens of 36 Australian passerine species, collected across their range from 1970 to 2012. In fact, they found no simple correlation of bill size and summer temperatures. In addition to the problem of a large bill becoming a heat trap at very high temperatures, it is very probable that it is also a problem in a desert winter when heat conservation is significant.

However, they did demonstrate a strong correlation between summer humidity and bill size, especially at somewhat lower maximum temperatures (at higher temperatures than the bird's body temperature, the issue of heat gain via the bill kicked in again, so the large bill was no longer advantageous whatever the humidity). Once temperatures get to the dangerously high levels above body temperature, evaporative cooling again becomes the only option, despite its inefficiency in humid situations and danger of dehydration in dry ones. This means that birds in a warming world living in already stressful hot lands may have ever-decreasing windows of thermal regulation opportunity open to them.

Smit *et al.* (2016) observed the responses to heat of a large number of desert bird species in the Kalahari. They confirmed that panting is effective at losing heat in dry conditions, but at a considerable cost in water loss, so it pays to put it off as long as possible (i.e. until it gets hot enough to be essential). They found that birds were better able to defer panting until higher temperatures if they were smaller, so could better dissipate heat via their proportionately larger surface area, if they could avoid excess activity (i.e. foraging in particular) when it was hot. If they didn't drink much, so were obtaining most of their water from their food, it was even more essential to use panting as a last resort.

However, you'll not be surprised to hear that there are even more complications to a bill's thermoregulatory function than surface area. Inside is a system of chambers called conchae, whose size and complexity determine both heat exchange and moisture retention during breathing. Danner *et al.* (2017) carried out detailed analyses of these conchae in two subspecies of Song Sparrow in the United States: one from arid coastal dunes, the other from a more humid situation. They took CT scans and radiographs of the bills, and showed that birds living in the arid situation had larger and more complex conchae, especially towards the bill tip, for better moisture retention.

(all African), Helmeted flocks will happily mingle at watering holes or food sources. On the other hand, they will unite furiously to repel a predator. This group would have ambled to water, feeding as they went, as soon as they left their roost site to which they return every night (e.g. Martínez and Bonan 2017).

Soon after their departure, their place was taken, to my very great pleasure, by a flock of Chestnut-bellied Sandgrouse. This Family of mostly desert specialists is found throughout much of Africa, plus western, central and southern Asia. Superficially pigeon-like with streamlined bodies, unlike the guineafowl they fly hard and far. Again unlike the guineafowl, which sometimes seem determined to be the centre of attention, sandgrouse specialise in being unobtrusive, from their soft brown cryptic colouring to their habit of lying still when nervous and avoiding any conflict with their fellows, which might attract predators. They rely on small hard seeds, and must fly daily to water, like Zebra Finches and Budgerigars in Australia. Here they may gather in huge numbers – at Lake Kabara in Mali up to 50 000 Chestnut-bellied Sandgrouse have been reported (de Juana and

Boesman 2016). But my favourite story about sandgrouse concerns their remarkable provision of water for their chicks, which lie in a simple scrape nest on the baking ground.

In 1896, prominent English ornithologist and conservationist – the splendidly named Edmund Gustavus Bloomfield Meade-Waldo – reported that he'd seen a male Pin-tailed Sandgrouse walk into water, soak his belly feathers with water, and carry it to nearby chicks, which immediately sucked it up. Sadly, and surprisingly given his excellent track record, people refused to believe him and then dismissed others who were reporting the same phenomenon. It was not until 1960, long after his death, that the world became reluctantly convinced by the mounting evidence. The male sandgrouse's belly feathers are superbly adapted to the purpose. A gram of them can hold up to 20 mL of water: four times what a kitchen sponge can manage. The feather barbules are soft and coiled, forming a felt-like material that absorbs water by capillary action and enables him to carry up to 40 mL of water for tens of kilometres. When he arrives, he stands over the chicks, which suck the water from his belly. He might have to make up to three trips a day for them (de Juana 2017). I could only hope that some of these Waza birds were loading up to feed thirsty babies.

Shortly after that, a most unpleasant stomach bug brought me very low indeed and I have limited memories of the rest of my time in the Sahel, but that's how birding can go sometimes and I would prefer my memories of Waza to be of guineafowl and sandgrouse than what followed. Unlike the now fearful people of Waza, I have a choice.

The Atacama: the driest desert

The mighty Atacama in northern Chile and southern Peru is a desert like no other. Parts of it are the driest places on Earth, with some locations recording no rainfall at all in the 450 years of European records. Overall, the average rainfall is 15 mm per annum, but the coastal cities of Iquique and Arica average less than 3 mm a year. It has been subject to 'extreme hyperaridity' for at least 3 million years, making it the oldest consistently arid area on the planet. One spin-off of the aridity is that in some places there are mountain peaks over

6000 m above sea level with no snow cover, although much lower Andean peaks right on the equator do have permanent snow caps. The basis of this aridity is the presence of the cold Humboldt Current against the Pacific coast: winds blowing across the chilly sea pick up very little water. Moreover, the Andes form a barrier to moisture-bearing winds blowing west across the Amazon basin from the Atlantic, creating a rain shadow. The west coasts of Australia and southern Africa also have deserts right to the sea due to similar cold currents, but even the forbidding Namib Desert in Namibia can scarcely compare to the Atacama. Near the sea, the camanchaca – the pall-like morning sea mist – can bring enough moisture for some tenacious plants to survive. Higher in the Andes, just enough moisture leaks over the ridges from the Amazon for some growth, mostly in the form of tough hard cushion plants. In between, from 1000 to 2000 m above sea level, is the 'absolute desert' where the zones of 'has never rained' lie. Here you can stare out of the vehicle window at a landscape of sand and pebbles that stretches to the horizon, uninterrupted by green.

Flamingos

Yet, counter-intuitively, there are still birds in the Atacama. Flamingos for instance ... Now, although I may well be losing my remaining marbles, the evidence is not in that observation. Yes, there are flamingos – three species in fact – that live and breed in the mighty Salar de Atacama, comprising 3000 km^2 of salt flats in a basin with no outflow in the mid-level Andes of Chile, near the desert town of San Pedro de Atacama. Water runs down from the soaring, snow-capped peaks that loom jaggedly into a sharp blue sky to the east. There is not much water, but enough to form shallow trapped lagoons super-saturated with salt. In these live the remarkable brine shrimps, which flamingos can't resist.

Laguna Chaxa is a surreal world. Fabian's Lizards peep out from little salt caves and splash through water that our feet would find caustic in pursuit of the myriad flies that swarm across the surface. Chilean, Andean and Puna (or James') Flamingos, Andean Avocets and Puna Plovers are among the mouth-watering birds (from a birding

The taxonomy of flamingos

Flamingos belong to an ancient lineage, so old that their relationships to other bird groups are unclear, though some smart money is currently on a connection with grebes. Once there were many species scattered throughout the world, including Australia, until just a couple of million years ago when the vast inland lakes dried out, but now there are just six species (or perhaps only five, depending on who you ask). The Greater Flamingo (of Africa, southern Europe and southern Asia), the American Flamingo (from the Caribbean and the Galápagos) and the Chilean Flamingo (from central Peru south along the Andes to Tierra del Fuego and east to the Atlantic) form a close species group. The Lesser Flamingo of Africa is given its own genus, and the Puna and Andean Flamingos, both Andean specialists, form a third genus.

point of view!) that work the shallow waters. All around, where the water has evaporated, is a plain of churned-up mud sparkling with salt crystals. And beyond, the Andes, snow-splashed above completely arid brown slopes.

The Laguna Chaxa flamingos are doing what is normal to them, but extraordinary to us. The head is upside down between the bird's feet, the horizontal inverted bill pumping water in and out. The large, fatty, highly sensitive tongue with numerous fleshy spines is the pump, working at up to 20 beats a minute, sucking in particulate-laden water and expelling unwanted items via a complex set of movements. This diet makes their feathers pink too, from the β-carotenes contained by the shrimps (see Photo 4). Chicks begin life grey and 'pink up' as they eat their shrimps. 'Pink' doesn't really do it justice, however, because flamingos flaunt a rich range of red-pink-orange – the ancient Egyptians used a flamingo-shaped hieroglyph to indicate red. (To keep them coloured, zoo flamingos are fed canthaxanthin, a naturally occurring β-carotene that is also used in tanning pills, mostly illegally.) All three of the Salar de Atacama species breed, often together, in colonies of cone-shaped mud nests topped with a bowl that holds one egg. Feeding the chicks involves another very singular flamingo adaptation – they produce milk! Well, they don't really, of course – that is definitively a mammalian characteristic – but, unlike almost all other birds, they synthesise a protein to give their chicks a boost. The protein is derived from glands throughout the upper digestive tract. Also unlike mammals, both flamingo sexes produce it and, among other birds, only pigeons and male Emperor Penguins have duplicated the trick.

After a few hours among the flamingos in this driest of deserts, driving back in the evening into the bustling oasis town of San Pedro de Atacama, with its green trees and rows of gringo-oriented souvenir stalls, seems stranger still.

Other memories of deserts

Like sand grains picked up by sudden breezes passing over desert dunes, some other desert memories drift into my mind:

- Waking in the Botswanan Kalahari Desert to a pair of Southern Yellow-billed Hornbills catching the earliest light, then giving a brief wing-stretching, bowing display, bubbling quietly the while.
- Okaukeujo Camp in Etosha National Park, based on a vast dry salt pan in north-western Namibia. The camp ground is dominated by huge Sociable Weaver nests, which can contain up to a tonne of grass and be metres long and high. The tiny neat black and white spotted architects are everywhere. From the top of the nest peer a couple of tenants: an exquisite chestnut, grey and white pair of Pygmy Falcons, just 20 cm long. Although they mostly eat out, they have been known to snack on the landlords' chicks.
- A magnificent Black-breasted Buzzard, chestnut and dark chocolate, drifting along the face of mighty red Uluru (for a while known as Ayer's Rock) in central Australia.
- Sitting on a brick-red sand dune near Cameron's Corner where New South Wales, Queensland and South Australia meet, watching, for the only time in my life, a family of white-streaked rusty little Eyrean Grasswrens bouncing about like ping pong balls between tussocks of cane grass. In all the world, they live only on dune crests in the Lake Eyre basin in the central Australian deserts.

References

Albright TP, Mutiibwa D, Gersond AR, Krabbe Smith E, Talbot WA, O'Neill JJ, *et al.* (2017) Mapping evaporative water loss in desert passerines reveals an expanding threat of lethal dehydration. *Proceedings of the National Academy of Sciences of the United States of America* **114**(9), 2283–2288. doi:10.1073/pnas.1613625114

Chisholm A (1958) *Bird Wonders of Australia.* Angus and Robertson, Sydney.

Danner RM, Gulson-Castillo ER, James HF, Dzielski SA, Frank DC, Sibbald ET, *et al.* (2017) Habitat-specific divergence of air conditioning structures in bird bills. *The Auk* **134**, 65–75. doi:10.1642/AUK-16-107.1

de Juana E (2017) Sandgrouse (*Pteroclidae*). In *Handbook of the Birds of the World Alive.* (Eds J del Hoyo, A Elliott, J Sargatal, DA Christie and E de Juana). Lynx Edicions, Barcelona, Spain, <http://www.hbw.com/node/52253>.

de Juana E, Boesman P (2016) Chestnut-bellied Sandgrouse (*Pterocles exustus*). In *Handbook of the Birds of the World Alive.* (Eds J del Hoyo, A Elliott, J Sargatal, DA Christie and E de Juana). Lynx Edicions, Barcelona, Spain, <http://www.hbw.com/node/54085>.

Elphick J (2014) *The World of Birds.* CSIRO Publishing, Melbourne.

Forshaw J (1994) The little Aussie breeder: budgerigars. *The Australian Geographer* **16**(2), 116–124.

Forshaw J, Shephard M, Pridham A (2012) *Grassfinches in Australia.* CSIRO Publishing, Melbourne.

Gardner JL, Symonds MRE, Joseph L, Ikin K, Stein J, Kruuk LEB (2016) Spatial variation in avian bill size is associated with humidity in summer among Australian passerines. *Climate Change Responses* **3**, 11. doi:10.1186/s40665-016-0026-z

Joseph L, Wilke T, Ten Have J, Chesser RT (2006) Implications of mitochondrial DNA polyphyly in two ecologically undifferentiated but morphologically distinct migratory birds, the Masked and White-Browed Woodswallows *Artamus* spp. of inland Australia. *Journal of Avian Biology* **37**, 625–636. doi:10.1111/j.0908-8857.2006.03767.x

Joseph L (2009) Woodswallow; a longer term, evolutionary view of boom and bust. In *Boom and Bust: Bird Stories for a Dry Country.* (Eds L Robin, R Heinsohn and L Joseph) pp. 205–222. CSIRO Publishing, Melbourne.

Martínez I, Bonan A (2017) Guineafowl (Numididae). In *Handbook of the Birds of the World Alive.* (Eds J del Hoyo, A Elliott, J Sargatal, DA Christie and E de Juana). Lynx Edicions, Barcelona, Spain, <http://www.hbw.com/node/52222>.

McDonald JD (1973) *Birds of Australia.* Reed, Sydney.

McEvey A (1976) Osteological notes on Grallinidae, Cracticidae and Artamidae. In *Proceedings of the 16th International Ornithological Congress.* 12–17 August 1974, Canberra (Eds HJ Frith and JJ Calaby) pp. 150–160. Australian Academy of Science, Canberra.

Morton S (2009) Rain and grass: lessons in how be a Zebra Finch. In *Boom and Bust: Bird Stories for a Dry Country.* (Eds L Robin, R Heinsohn and L Joseph) pp. 45–73. CSIRO Publishing, Melbourne.

Payne R (2017) Australian Zebra Finch (*Taeniopygia castanotis*). In *Handbook of the Birds of the World Alive.* (Eds J del Hoyo, A Elliott, J Sargatal, DA Christie and E de Juana). Lynx Edicions, Barcelona, Spain, <http://www.hbw.com/node/61171>.

Payne R, Bonan A (2017) Waxbills (Estrildidae). In *Handbook of the Birds of the World Alive.* (Eds J del Hoyo, A Elliott, J Sargatal, DA Christie, E de Juana). Lynx Edicions, Barcelona, Spain, <http://www.hbw.com/node/52373>.

Rentz D (1991) Orthoptera. In *The Insects of Australia*. 2nd edn, Vol. 1. CSIRO Division of Entomology. Melbourne University Press, Melbourne.

Smit B, Zietsman G, Martin RO, Cunningham SJA, McKechnie E, Hockey PAR (2016) Behavioural responses to heat in desert birds: implications for predicting vulnerability to climate warming. *Climate Change Responses* **3**, 9. doi:10.1186/s40665-016-0023-2

Symmons P (1985) Locusts: the plague of '84. *Australian Natural History* **21**(8), 327–330.

Tatersall GJ, Andrade DV, Abe AS (2009) Heat exchange from the toucan bill reveals a controllable vascular thermal radiator. *Science* **325**(5939), 468–470.

USGS (2016) *What is a Desert?* United States Geological Survey, Reston VA, USA, <https://pubs.usgs.gov/gip/deserts/what/>.

van Dyck S (1995) Budgerigars: mini Australian megastars. *Australian Natural History* **24**(12), 20–21.

Zann RA, Straw B (1984) Feeding ecology and breeding of zebra finches in farmland in northern Victoria. *Australian Wildlife Research* **11**, 533–552. doi:10.1071/WR9840533

2
Rainforests

Western Cameroon: lowland tropical rainforest

Early April, the beginning of the rainy season in Korup National Park: a steaming dark rampantly green world, tangled with vines. Fungi sprouted everywhere, helping turn trees and butterflies to soil and back again to trees and butterflies. There were fungal copses of tiny white and orange buttons, crops of yellow corrugated cups on logs, mauve parasols, huge vermilion and tiny pearly white brackets, among so many others. Columnar termite mounds wore mossy domed caps, like slightly phallic gnomes. Korup has a long bird list, but they were remarkably inconspicuous. We could hear them but even seeing canopy movement, let alone identifying it, was a rare event. Every now and then the air would vibrate with the throbbing 'whop' of one of the great casqued hornbills flying over, but the canopy might as well have been a ceiling. Rainforest birding isn't always that tough, but it can be.

I had always wanted to see a tropical African rainforest and the mighty Korup is an extraordinary place. It adjoins Nigeria on the western border of Cameroon, which is in the corner of Africa, where the north-running west coast of the southern part of the continent suddenly swings at right angles to the west. Korup's 125 000 ha of lowland forest is of world significance and in 1986 the Cameroon Government declared it the nation's first national park (see Photo 5). We – a small international group of birdwatchers, the same ones you met briefly at Waza – had swayed across the enormous Mana River suspension bridge (in my case caught between a deep desire not to look down, and a need to do so to avoid gaps in the planks) and walked 8 km into our camp site. After a couple of minutes, we'd passed a sign advising that Prince Charles had 'trekked' to here in 1990.

A rainforest requires, unsurprisingly, rain. In fact it requires a lot of rain (a minimum of 2500 mm a year, by most definitions) and moreover it has to be constant enough so that there is no really dry season. Tropical rainforests are characterised by a closed tree canopy and a desperate struggle for light, leading to towering trees and smaller plants, vines and epiphytes, which use the trees to get out of the understorey's gloom. Many rainforest trees have buttresses, whose advantage to the tree is still debated, and such few plants as can grow on the dark forest floor have leaves of deepest green, packed with chlorophyll to enable photosynthesis down where only 1% of the light on the canopy penetrates. Up there it's a different world: temperatures fluctuate from 40°C during the day to mid-20s at night, and humidity from 90% at night to 60% on a sunny day. Strong winds and wild storms are prevalent. Down where we are, there is no wind, the temperature and 90% humidity scarcely vary over 24 h. Tropical rainforests are found throughout the low latitudes – though clearing is progressing at a horrific rate – from sea level to high in the mountains, though there are different types of rainforest recognised in different situations.

Late in the afternoon, we left our Korup camp to walk another couple of kilometres to a site that to me was literally the stuff of dreams. I don't remember exactly when I first read about Picathartes ('magpie vulture'), but it was well over 40 years ago and I suspect via either Gerald Durrell or David Attenborough, back in another world when ripping animals from the wild and incarcerating them in a faraway zoo seemed a reasonable thing to do. Nor do I remember why this bird had grabbed the attention of a boy who, though passionate about animals, was mostly obsessed by what my current guides would call 'hairy and scary'.

When birding is tough: Picathartes

There are just two members of the family, both endemic to West Africa and also known as rockfowl. The White-necked Picathartes is found to the west, from Ghana to Guinea; 'our' species, the Grey-necked Picathartes, lives around the 'elbow' of Africa, in Cameroon, Gabon and Equatorial Guinea. Both are large passerines, about the weight of an Australian Magpie. Their relationship with other bird groups remains entirely unclear, and the birds are rarely encountered away from their nesting sites. These tend to be colonial, with big mud nests on cliffs that are angled at 70–80° so that they provide shelter from the rain. For reasons that are not understood, the birds visit the nests at dusk and dawn even out of breeding season and this is what we were relying on.

We had intended to bird on the way, but the storm clouds were so stygian that it was hard to see our feet in the forest, let alone any putative birds. Korup has an average annual rainfall of over 5 m, and it didn't disappoint. Curiously the light increased again as the rain started and it remained at the same slightly higher level until sunset. I must confess to feeling a bit negative about it all – it had been a gruelling trip and I was weary; we hadn't seen many other birds in the forest, so why would we see this very difficult one, especially as it was going to be dark and wet? We sat down, in my case across a sharp ridge of granite, and were exhorted not to move. And then something wonderful happened. In the steady rain – at some stage it ceased falling from the clouds but continued unabated from the foliage, as is the way in rainforests – my physical discomfort receded as I took in the beautiful pale granite cliff face and absorbed what was happening. I was actually sitting in Korup and the giant swallow-like nests I was looking at belonged to Picathartes; how dare I be anything but awed at such a privilege? This was very special and suddenly it didn't matter if the bird came or not.

But it did come. A swoop across the cliff face, a quick inspection of the nest and down to the ground, all a blur, then up onto a low rock ledge. Perfectly framed for me by two trees, this lanky grey bird with gleaming bare head, brilliant red crown and blue forehead, seemed to peer straight at me. For five seconds, ten? I have no idea; the real answer though is 'forever', for the image is still bright and sharp in my mind and will be so throughout my life. As a boy I never dreamt I would see a live Picathartes, or any of those wonderful animals that dwelt only in the magical world of the books introduced to me by my father. The only specific thought I can recall on the way back to camp was regret that I would not be able to tell him that this dream he'd triggered had come to life. Beyond that, I was just dripping wet and euphoric – and I knew that neither birding, nor much else that I could think of, got better than this.

When birding is tough: Mt Kupé Bushshrike

Rainforest birding, however, is not always euphoria-inducing. The village of Nyasoso in mountainous western Cameroon is reached via a ferocious drive from Tombel (home of the modest little office of the National Association of Snail Farmers of Cameroon) over potholes with bits of road attached. We stayed at the Nyasoso Women's Collective rooms: mine was a basic concrete cell with a metal single bed, but it did surprisingly have an en suite (cold) shower and toilet. Unfortunately, the shower drain hole blocked up and after a shower my bed stood ankle deep in a large pool of water, which had nowhere else to go. Remarkably, my bag happened to be on the high part of the floor!

Our target was one of the rarest and most threatened birds in the world, the Mount Kupé Bushshrike, with a population of perhaps only 50 pairs in its Bakossi Mountains stronghold. It was believed until recently to be restricted to here, though there are known now also to be some in neighbouring south-eastern Nigeria.

We began the day with a 3.45 am breakfast consisting of dry cereal on a flat plate, white bread and tea. At 4 am 10 of us piled into a short wheel base Toyota; four of us were in the 'back back', two each along the sides, facing each other on 20 cm wide metal benches, vaguely padded with a strip of rotten foam. Our knees were jammed against those of the person sitting opposite. It was a long 90 minutes to the village at the end of the road, on a very rudimentary steep bush track.

In the pre-dawn, we were greeted individually by Chief Abwé, a slight old man in a wooden shack only a little larger than those of his neighbours. It was too dark to see the nature of the photos in the newspaper clippings on the walls. Mercifully we were spared the traditional beer-drinking ceremony, though I'm not sure that our guides were.

We had barely set out when, at a creek crossing where three women already washing clothes accepted our intrusion very graciously, we had an unpleasant interruption. A small group of agitated and angry young men overtook us and blocked our way. Their leader, apparently badly affected by alcohol or drugs, was bizarrely still using a pink toothbrush, which somehow increased the sense of menace. A crowd materialised, with most concurring with the chap who told the angry ones, 'If you are wise, my friends, you will leave'. Unfortunately, they weren't and didn't. In the end it was all very tawdry, and came down to more money, which they felt their chief was not distributing properly; clearly the tradition of respect for the chief was eroding. They didn't want the cash for the benefit of the village though – they were quite open in explaining that it was for cheap whisky, and the extra 40 000 Central African Francs (about A$100) they eventually received would have bought them a lot of it. It would also have paid for a lot of food or school books.

It is always stimulating to be in primary rainforest, but when you are focussed on a particular target the birds can seem very sparse – I saw 14 species all day. A steep climb was the precursor to a hideous descent on a scarcely present track, evil-slippery with mud and leaves. Without trackside saplings, it would have been very dangerous indeed. We spent an hour listening and intermittently tape-playing its call up on the next plateau, but nothing. We scrambled, already wearily, down the other side, where someone thought they heard it along the valley so we hauled back up again and repeated the exercise, for the same result. Leaving even the rudimentary track, we then descended into the valley and battled up the other side – and then back again. My notes simply say 'this was *bad'*. I've done my share of bush-bashing, and was at about the median age of the group, but I recorded that 'I felt I was at my limits'. In the last hour of the final descent, on already perilously greasy slopes, the inevitable rain exacerbated the conditions.

We never did see the bird, not that day, nor the next, despite a brutal ascent (and descent!) of nearby Mount Kupé, which looms over Nyasoso. By then I was too sore to care much. My state may be judged by my reaction to a somewhat surreal encounter on my way down Mount Kupé, in late afternoon with thunder building. Most of the group had gone down earlier; I'd stayed with the hard core, but when, during the eventual and reluctant descent, they yet again turned back up-slope in response to a note that may have emanated from a bush-shrike I called it a day and continued down. The group leader wanted to make sure I was OK and sent Albert, our local guide, down after me. I discovered this only when I suddenly became aware of a figure bounding down the track towards me, brandishing a machete and yelling, 'I am coming for you!'. My reaction, recorded that night, was just, 'this will be interesting'.

Part of the essence of birding is the uncertainty, and my life has continued quite happily without a Mount Kupé Bushshrike ever being part of it. I have no yen to return to the Bakossi Mountains, but I certainly don't regret having been there. If the sole point had been finding the bird (as it was for at least a couple of my companions), then as it turned out there would have been no point. But the older I get, the more I am sure that the most important thing, in birding as in life, is looking *for*: looking *at* is just a bonus, never to be presumed.

Julatten, north Queensland: a magnet for birders

Traditionally it is deserts that are reputed to set us thinking, but I've caught some worthwhile trains of thought from a rainforest platform too. I was camping once at the incomparable Kingfisher Park near Julatten on the tablelands above Cairns in tropical Queensland. This property contains remnant rainforest, an extensive tropical orchard and accommodation specifically geared towards birdwatchers. I was woken with a shock in the dark small hours by a quavering nearby scream of 'he.e.e.elp' and was initially frozen rigid. My muscles gradually unlocked as I realised it was one of the more or less uniformly bizarre calls of the Orange-footed Scrubfowl: small incubating-mound builders of the rainforests and monsoon forests of Australia, New Guinea and associated islands. Doubtless my nightmarish awakening was fuelled by the murder mystery I'd been reading before going to sleep.

Mount Kinabalu, Sabah

The rich primary rainforest at Sepilok in eastern Sabah, Malaysian Borneo, is where orphaned Bornean Orangutans are rehabilitated and taught how to live in the wild. Not far from Sepilok is mighty Mount

Birds and death

We have long embroiled birds in our own complex belief systems surrounding death. Any sentence that contains 'humans', 'birds' and 'death' is more likely to be an unhappy one for the birds than us but, even when the subject of it is human death, we have often managed to hold the poor birds responsible for it! Unusual bird calls in the night are always likely to engender responses based on fear and ignorance, and doubtless the grim reputations often associated with owls are related to that as well, especially considering the hoarse scream of the Barn Owl. (Living in church towers, being basically white and hunting over cemeteries would have helped there too.) Surprisingly, it seems that this association of owls with death is near universal. Even now, they are believed to presage death and illness in parts of Mexico, Central America and the Caribbean, Argentina, Arab cultures, China, the Mediterranean, south, north and west Africa and Madagascar (Marks *et al.* 2016). Whether we can join Finnish owl expert Heimo Mikkola in asserting that this is because such beliefs came out of Africa with our ancestors is another question (Mikkola 2014). It seems to me that the forward-facing eyes of owls, enabling binocular vision, which is denied to most other birds, have been recognised as giving a human-like visage, bringing associations with both wisdom and evil. It's a fair assessment of us, but a bit rough on the owl.

Lapwings, migrating geese, curlews and whimbrels all call as they fly overhead at night in Europe, and some of their calls are very evocative indeed. It is no coincidence that these birds are intricately part of one of the most fearsome of English myths, the Seven Whistlers, which fly calling at night. Hearing six is bad news and death will follow, but if the six ever meet the seventh, the world will end. They are sometimes regarded as synonymous with the Gabriel Hounds (i.e. 'corpse hounds' apparently), or the Wild Hunt, which probably originated in Norse mythology and which go by many names in Britain and doubtless elsewhere. Along the way, they segued neatly into Christian folklore and added the Devil as hunt-master, without breaking formation. All agree that to hear them is very bad, but to see them is death (e.g. Simpson and Roud 2003).

Many stories revolve around people becoming birds on dying, probably the most famous in our culture being the belief, whose origin I cannot determine, among European sailors that their souls transform to albatrosses after death. That one seemed to protect the birds too, at least until long-line fishing was invented.

Kinabalu, surely one of the world's most imposing mountains. From a distance, it sprawls across the skyline with a mass of jagged granite teeth snapping at the clouds. The highest point is 4100 m above sea level. Most of the slopes are cloaked in dark green rainforest (except where the landslips that followed the totally unexpected earthquake of

2015 have ripped long pale gashes through it). I've never climbed to the top – I have no desire to climb for the sake of it – but have spent memorable time in the lower to mid forests.

Giant mosses and carnivorous pitcher plants vie for attention with the birds – and all provide an excuse for frequent stops on the relentless climb, with sets of steps rising one after the other. A pair of appropriately named Crimson-headed Partridges scurried across the track in front of us while equally red Temminck's Sunbirds worked the flowers above. As huge front-heavy Rhinoceros Hornbills flew over and busy little Plain Pygmy Squirrels scurried in the undergrowth, a richly coloured pair of Maroon Woodpeckers attacked the rotting end of a broken branch above our heads. They were too engrossed to treat us to their raucous call on this occasion, a call which used to be welcomed by local Dayak raiding parties, because it pre-echoed their own enthusiasm when they relieved one of the opposition of his head (Winkler and Christie 2016). Flocks of pretty little Chestnut-headed Yuhinas, grey above and white below with rusty head, assiduously searched leaves for insects as they passed quickly through. As we climbed higher, Mountain Blackeyes appeared (relatively large green white-eyes with an incongruous black eyepatch), probing their long thin bill into crevices in lichen-enshrouded branches. By the time we stopped for lunch at a small shelter, where Mountain Ground Squirrels and Mountain Treeshrews hung hopefully about, the clouds were descending until we were in a world of mist-wraiths. The mood suddenly grew sombre when a crew of rescuers hurried down the track past us with a still form on a stretcher. More and more walkers crowded into the shelter as the rain exploded onto the forest and our flimsy roof. Every now and then individuals or small groups would accept that it wasn't stopping any time soon and headed out into it; we eventually did so too, heading down a track that was now a torrent, water rushing over the tops of our boots. The pretty little waterfall near the bottom of the walk was now a roaring foaming maelstrom.

Back at the entrance station, the rain started to ease off and we were able to retrieve our binoculars from dry-bags in back packs. An immediate reward was a beautiful Golden-naped Barbet scoffing

Red, green and blue: tricky colours for feathers

These are interesting colours for feathers. The bright red features of the partridge and sunbird we had enjoyed before the rain at Kinabalu are due to the presence of pigments called carotenoids – there are many different ones. Curiously, most birds can't manufacture carotenoids but must obtain them from other sources, such as plants, algae or bacteria, or animals that have already eaten them. This is energy-demanding, extracting them from food and moving them to where they are required in the body. But it gets worse: red carotenoids are hard to come by (flamingos get them from shrimps that have eaten carotenoid-rich algae), so most birds that want them must eat yellow carotenoids then convert them to red, at an even higher energy cost. Presumably the colour is a statement of the owner's fitness: that he can afford to squander energy simply in order to look gorgeous, and hopefully she'll be impressed enough to mate with him. In many red species, only the male indulges, but parrots have come up with a red pigment apparently unique to them, called psittacofulvin, which they manufacture themselves. Accordingly, parrot females are as likely as males to be red. The barbet has employed yellow carotenoids, with no conversion, for its lovely nape crescent.

Blue, however, is another story entirely. Despite the seemingly incontrovertible evidence of our eyes, there are no blue pigments in bird feathers. Instead it's all a clever use of structure to ensure that only blue light is reflected. Tiny air bubbles are embedded in feather barbules, of just the right size to absorb red and yellow light so they don't reflect, but blue light does. If you were to take a red feather from a sunbird and a blue one from a barbet and soak them in an appropriate solvent, the red one would fade to white but the blue one would still be blue. On the other hand, if you were to crush the feathers with a hammer or in a vice, the red would be unchanged but the blue one would now be colourless with the bubbles destroyed.

What about the green of the barbet? Nope, another trick, or rather the same one. Very few birds indeed have green pigments, despite the plethora of green birds, particularly among parrots. Like them, the barbet has yellow pigments across most of its body. Most of these yellow feathers, however (other than the obvious ones on the back of the neck), also have the same blue-reflecting bubbles that those of its face have – and of course blue mixed with yellow gives green (e.g. Hill 2010; Elphick 2014, pp. 79–83).

Bannerman's Turaco, like most of the 27 members of its family, all living in sub-Saharan Africa, provides a dramatic exception to the generalisations above about red and green feathers. It is a stunning bird, with metallic green wings slashed with red, a gleaming red nape and peaked crown, yellow bill and a glossy long dark blue tail. The red, however, is not based on carotenoids, and the green is purely due to pigments. Both red turacins and green turacoverdins are copper-based compounds, synthesised in the bird's body from copper ingested in fruit – it takes around 3 months for a turaco to eat enough fruit (20 kg, containing 20 mg of copper) to make its feathers so green and so red (e.g. Turner 2016). Until recently it was believed that these compounds were unique to turacos, but turacoverdins (or something very similar) are now known from a jacana, a pheasant and the Indonesian Crested Partridge whose colour patterns look remarkably like those of a turaco.

berries in a bush at eye level. Barbets are close relatives of toucans, but without the outlandish bill – chunky medium-sized fruit-eaters, mostly brightly coloured, from Africa, south and South-East Asia and the South American tropics. In Asia and South America they are largely rainforest birds. Until recently they were all lumped into one family, but now they are split into a family for each continent (plus one for the delightful two species of South and Central American toucan-barbets). Besides the eponymous nape, the Golden-naped was overall a brilliant leaf-green with a bright blue crown, forehead and throat (see Photo 6).

Bamenda Highlands, Cameroon: vanishing birds

Bannerman's Turaco has the misfortune to live only in a seemingly doomed habitat. South-western Cameroon comprises a bulge into Nigeria. The Bamenda Highlands, comprising the sole habitat of this turaco, lie in the north of this bulge. (For some reason they are classified as in 'north-west Cameroon'.) The Highlands comprise a volcanic plateau between 1000 and 2000 m above sea level, relatively cool and very wet, with a 9-month rainy season. Once they were heavily forested, but overpopulation, repeated burning to clear the forests for subsistence farming and ever-spreading eucalypt plantations for firewood have reduced the rich forests to tiny tendrils of green along streamlines: tendrils that are being gnawed away all the time. Bannerman's Turaco seems surprisingly capable of surviving in small forest remnants, but this ability is not infinite. The drive into the highlands from the town of Bamenda is profoundly depressing, through a countryside stripped bare of trees – other than eucalypts – even on hills too steep for any useful production. We located the site of a remnant that our guide had birded 2 years earlier, but it was now gone.

Finally, we moved to Lake Awing: a large natural lake, in the same lake system as Lake Nyos which, in 1986, emitted overnight a massive cloud of volcanically derived carbon dioxide, killing 1700 people. The lake is of considerable cultural significance locally, but is environmentally devastated, surrounded by only a narrow strip of vegetation with a couple of pitifully narrow forested gullies running off it. Here we found the turaco, along with the perky little black and white Banded Wattle-eye, which belongs to a Family of small robust African insect eaters. It too is restricted to the shrinking Bamenda forests. The very smartly turned out gold and black Bannerman's Weaver and the rusty-coloured Bangwa Forest Warbler fortunately also live in a couple of other sites, but their future is also at best uncertain.

Normally the opportunity to see special birds that live only in a small remote part of the world is a privilege that brings pleasure and excitement. This time, however, I left just feeling sad at the thought that these beautiful birds, which have evolved in this place and lived here for hundreds of thousands of years, could well have disappeared from the world before I do. And, of course, with them will go countless plants and small animals whose existence we're probably not even aware of. Perhaps if more birders were to visit, and if our money could stay in the villages, things might be different, but they're not. Ecotourism here isn't doing much harm, but it's not doing a lot of good either.

Mindo Valley, Ecuador: a jewel in the crown

In contrast, Ecuador is a guiding light in how bird-oriented tourism, especially in rainforests, can benefit communities and the environment. Ecuador has a National Avitourism Strategy! It is one of the most extraordinary places for birds (and thus birders!) in the world, with 16% of the world's bird species in 0.05% of the Earth's land area. It is perhaps even more striking to put it into the context of the remarkable bird wealth of South America: Ecuador hosts 48% of the continent's multitudinous species in just 1.6% of its area. Ecuador's finest jewel is the Mindo Valley and surrounds, part of the Chocó cloud forest bioregion, one of the world's biodiversity hotspots. It lies not far north-west of Quito on the far side of Pichincha Volcano, which looms over the national capital. Here everything seems geared towards birders –

Cloud forests

Cloud forests are generally found in the tropics or nearby, at higher elevations than many rainforests, in situations where, as the name suggests, cloud sits low on or below the canopy for at least part of most days, especially in the afternoon. The moisture in the cloud condenses on leaves and drips constantly to the floor. The exclusion of sunlight for much of the day means both that light levels are low (even by rainforest standards) and that evaporation is reduced, so that everything is always drenched. Ferns, mosses and lichens are abundant and it's easier to hear birds than to see them. They are there though – cloud forests are very rich and often feature a high proportion of endemic species.

we're worth a lot to the local economy and, as a result, the mist-swathed forests are fairly well cared for, and infinitely more so than in the Bamenda Highlands.

Paz de las Aves: positive ecotourism news

A pioneer in the rebirth of the misty Mindo Valley as a must-visit destination for the world's birders has been a quietly spoken, often-smiling former subsistence farmer called Angel Paz. At the turn of the century, he and his family were of necessity contributing in their small way to the incremental loss of cloud forest to small-scale farming. He was a visionary, however, and when he found a small Andean Cock-of-the-Rock lek (more of that anon) at the bottom of a steep forested gully, he built a track down to it to make it accessible to visitors, encouraged by local bird lodges. In the building process, a large plain bird began to follow him, hopping across the ground to harvest worms thrown up by the excavations. He knew you could eat it, but fortunately did not act on that knowledge. Local bird guides let him know that cocks-of-the-rock were 'birding silver' all right – but antpittas (including this Giant Antpitta) were pure gold! Members of the vast South American assemblage of ancient Gondwanan suboscine passerines, they are mythically hard to see in the dense rainforests and cloud forests where they dwell, causing heartache to thousands of birders. Yet here was one hopping boldly in the open …

Angel is a man of seemingly infinite patience and persistence, and with a remarkable knowledge of the forest and its birds. He knows all their calls and can seemingly carry on conversations with them – they certainly respond to his whistles in kind. With these attributes, over months of work, he found and habituated species after species of antpitta (all otherwise near-impossible for a visiting birder to see) to emerge to feed on wild-gathered forest earthworms. He tried to offer them worms from his compost heap, but they were spurned. I have now seen five species with him (Giant, Ochre-breasted, Yellow-breasted, Chestnut-crowned and Moustached, for the record) but it doesn't end there.

A flock of Dark-backed Wood Quail (New World quail, not at all related to Old World ones) comes from the forest to demolish bananas on a forest track. My book says they are encountered 'only by chance' – not here though. Tapaculos are another old South American group of uber-skulkers: total heart-breakers. The magnificently named Ocellated Tapaculo is a most striking bird, 22 cm long with white paint splashes on a black canvas, and rusty face and rump – or it would be striking if you could ever see it. It is one of *the* voices of the cloud forest: a falling whistle that punches through the mist, approaching but never quite arriving. At Paz de las Aves ('Peace of the Birds', as well as the play on Angel's family name) I sat in sheer disbelief as an Ocellated Tapaculo emerged from the forest to accept Angel's worms, a very few metres from my feet (see Photo 7). None of the birds are dependent on Angel's worms or fruit offerings, and Angel constantly changes feeding sites to alleviate the risk of predators taking advantage.

Angel, assisted by his brother Rodrigo and more recently, I have read, his son Venicio, has persisted in the face of initial opposition from his wife (who thought he could have been doing more productive things than feeding his forest chooks) and scorn from neighbours and some family members. No-one now suggests he wasted his time. They have built a nice-looking guest house on site, where people can stay and enjoy local produce, instead of making the pre-dawn drive from Mindo. The family is now comfortably off and have diversified into fruit growing, which has inevitably attracted more birds.

This is a good news story for the family, for the birds and for visiting birders, but it is a lot more than that. Many people, both locally and further afield, have seen that looking after birds – which, of course, means their habitats – makes good financial sense, as well as the acceptance of a custodial responsibility. If that rainforest down in the gully might be worth money, it makes more sense to leave it alone than to clear it for a few more banana plants. Angel has been invited to train people in private and government reserves to habituate some of their local 'difficult' birds in other parts of Ecuador and Colombia; in turn, more people will visit there and spend money locally, and the message spreads.

Moreover, it's harder to get people, locals or visitors, to care about a bird they never see. Familiarity can engender a sense of connection and care, which can lead to that essential assumption of the mantle of custodianship.

Aguas Verdes, northern Peru: more positive ecotourism news

Further south, in the north of Peru but across on the eastern slopes of the Andes (the opposite side from Angel's property) in the village of Aguas Verdes, I visited another new venture in rainforest conservation by a farming family. Although I didn't ask them, I'm sure they haven't been taught by Angel, because their approach to feeding forest birds is different and they focus on different bird groups. Norbil Becerra is a carpenter who was intending to turn his small family-owned patch of rainforest just outside the village into a coffee plantation. These forests at 1500 m above sea level support a mixture of higher elevation cloud forest species and lower level rainforest birds.

However, just in time, Norbil visited nearby Huembo Lodge, run by ECOAN, the Asociación Ecosistemas Andinos (Association of Andean Ecosystems): a research and conservation foundation. There he saw his very first hummingbird, which just happened to be the extraordinary Marvellous Spatuletail (and no group of birds has a more delightfully eclectic collection of names than the sublime hummers). Marvellous indeed, but more than that: 'ridiculously unlikely' springs to mind. He has a glittering blue crown, an iridescent blue/green/purple throat, green back and white belly with a black line down it. So far, just another hummingbird – which is to say, superb. However, then there's the tail … The tail feathers are reduced to just two pairs. The central pair are long and spine-like, but the outer pair are immensely long –

some three times the length of his body – and crossed over. Most of their length comprises bare shaft, but they are tipped with purple 'racquets' of feather, which wave as he dances on air. I'd be astonished if Norbil hadn't been smitten by them.

Back home, he abandoned plans for the coffee plantation and instead applied his trade to building a truly magnificent raised and covered viewing platform looking out into flowering plantings, selected to attract hummingbirds and butterflies. Norbil is clearly a special person and his determination and resilience in the face of pressure are remarkable. It took 7 months for the recalcitrant hummers to find the feeders, but every day he cleaned and refilled the feeders with sugar solution while the pressures grew around him to be 'sensible' and plant the coffee crop instead. In the end the hummers came, and now visitors are coming too. I was there in late 2015, about a year after it opened. We arrived in the village in a torrential rainstorm and were invited into a simple two-storey home, with just one open downstairs room (through which a very well-behaved pig wandered en route from the street to the back yard), to drink tea while we waited for the skies to relent. Eventually we were led along the still-streaming unsealed street and along a road out of town for about a kilometre. The white sands underlying the forest drain the water away very efficiently.

We were rewarded with an enthralling array of hummers, including some restricted range ones that were new to me. The stand-out for me, however, was the amazing Wire-crested Thorntail, with long forked tail and sparkling green throat and forehead feathers that continue upward into a long slender crest (see Photo 8). As it performed its magic in hovering at the pink verbena flowers just in front of us, I was very grateful to Norbil indeed.

So far it's a good story, but not unique – many places in the northern Andes put out hummingbird feeders, though mostly they are associated with lodges. However, Norbil then tried something else. In the forest, just off the access track to the hummingbird viewing platform, he built a small raised hide with viewing slots facing the forest floor just a few metres away. A pipe delivers corn from the hide to the field of view provided by the slots. And birds have already learnt to come for the

Tinamous

Tinamous fascinate me. They are among the most ancient of living birds, some 50 species restricted to Central and South America. Superficially like big quail, until recently there was some disagreement as to whether they belong in the same Order of birds as the ratites, which are flightless and mostly giant old Gondwanans such as the ostriches, emus, cassowaries, kiwis and rheas, or in an associated Order. Recently, however, the argument seems to have been settled in favour of the former (see pages 184–5). Tinamous, however, have retained their powers of flight, though they fly only weakly and awkwardly for the most part. This limitation is due to several factors, including small wings, the lack of a tail to help with braking and steering, and a surprisingly small heart. All birds are dinosaurs, but tinamous and other ratites have retained some particularly reptilian traits, such as their blood proteins and the structure of their palate, which closely resembles that of *Tyrannosaurus rex*. Repeated unsuccessful attempts have been made to introduce various species to Germany, France, Hungary and the USA, so that people can enjoy the fun of killing them. Curious creatures – us I mean, not the tinamous.

corn, including two species of tinamous (Cinereous and Little) while we were there. I had almost despaired of seeing wild tinamous, most of which lurk in the gloom of the forest floor, yet here they were pottering about right in front of me.

Other hard-to-see forest floor birds that came that day to Norbil's hide included Orange-billed Sparrows, Grey-necked Wood Rails and White-tipped Doves.

If the only way to see birds was by such habituation, the world would have problems well beyond limits on our birding, but clearly that is not the case for most species, and it's probable that such approaches wouldn't work for many species anyway. Norbil's birds, like Angel's, are not dependent on the handouts, though they doubtless benefit from them. They most certainly benefit from the fact that their forest is still there and not converted to a coffee crop. Moreover, as with Angel's neighbours in Mindo, it is likely that at least some of Norbil's neighbours, as well as others further afield, will be inspired to do something similar. Their lives will materially improve and yet others will see that it is due to protecting the forest and its birds. It's hard to see any losers in this scenario.

Acjanaco Pass Road, southern Peru: cloud forest

My introduction to neotropical rainforests was at the other end of the Peruvian Andes, far to the south, crossing the watershed of the great range north-east of the old Inca capital of Cusco, at Acjanaco Pass, 3500 m above sea level. Here we are in real cloud forest: a stunted high-altitude version also known as elfin forest. It's a strange and wonderful world, where ghostly streamers might dissipate nebulously, or clarify into ribbons of lichen or moss. Birds appear from the mist, perch briefly and disappear into it again. Giant Hummingbirds, Great Thrushes, Masked Flowerpiercers, Grey-breasted Mountain Toucans all come and go, often in silence. There is a wealth of orchids on the trunks and branches, which also coalesce out of the swirling clouds as we walk among the trees. Behind us to the west from where we came are drier slopes in a rain shadow; ahead of us are lush forests, falling to the vast Amazon basin. The road continues in that direction, descending through changing cloud forest through an endless series of blind switchbacks and along sheer drops into the green depths, past waterfalls and rockfalls. Some 2000 m below Acjanaco, we stopped and walked briefly off the road to a hide – in front of which, inconceivably as it seemed to me, was a group of some of the most outrageous-looking birds on the planet, and another species that I never really expected to see.

Andean Cock-of-the-Rock: leks

The Andean Cock-of-the-Rock is like no other bird. It's a large cotinga: a group of the numerous old South American suboscine passerines, about the size of a domestic fowl. The female is a uniform coppery brown; the male is definitely not! His head and body are bright red here, but orange in populations north of Cusco, both shades being classic examples of showy and expensive red carotenoids. He has a large pearly-grey patch across his lower back formed by the tertial feathers (the wing feathers closest to the body), which cover and protect the flight feathers when they are closed, while the rest of his wings and tail are black. A broad comb-like red crest on his head makes his whole head look ridiculously long, and apparently half way along it the bill barely protrudes from the mass of feathers. His eyes with black pupil and pale yellow iris resemble those of a teddy bear. And then there's the behaviour … Pairs of males display to each other, bobbing heads, clicking bills, leaping about and flapping wings, while emitting a hoarse buzzing squawk. When a female appears, the displays rise in intensity to a crescendo. This was not only my first experience of this bird, but of a lek.

The word lek is from Swedish, and was coined to describe the display areas and gatherings of various grouse species. These gatherings comprise dozens of males that have come together to flaunt their often outlandish garb and make themselves loud and conspicuous. (In fact it's all somewhat reminiscent of certain social situations involving young human males, and the reasons for both are rather similar.) In many leks – perhaps even most, including some human ones – most of the females' selection process is done for them, as the most successful males work their way via a series of contests to the centre of the display area. It is no coincidence that these positions are the safest, with predators inevitably drawn to the edges of the performance. It is certainly to the females' benefit not to have to travel far to compare the talent on offer, and for said talent to sort out their rankings for her to inspect. The benefit to successful males is also clear, with guaranteed access to as many females as are impressed by the overall performance of the lek. Unlike in other mating systems, however, the less successful blokes, the majority of the male population, will probably miss out entirely, even if they don't get picked off by lurking predators (e.g. Fiske *et al.* 1998).

In addition to grouse and cotingas, lekking is practised by some members of groups of animals as disparate as hummingbirds, African antelopes, Galápagos Marine Iguanas, frogs, cicadas and butterflies (where the term used is 'hill-topping'). A variant of lekking is the 'exploded lek' where competing males are distant from each other and compete by calling loudly; examples include the tragically endangered flightless New Zealand Kakapo (a parrot), and the Australian Musk Duck, whose splashes and whistles and grunts carry far across the surface of the water (see page 130).

In the case of the cocks-of-the-rock, a series of one-on-one knockout competitions sees consecutive winners take up the prized innermost perches and, once a female has made up her mind, she simply ignores the also-rans and goes straight to the top. Life, however, is never simple and straightforward (for humans or cotingas). His frantic displays, on top of his high-energy red carotenoid habit, sap his strength – but if he goes for a much-needed fruit snack, he loses his prized and hard-won position and must start all over again. The result for enthralled birders in the hide is that, no matter how rapt we are (and it is hard not to be), eventually we are going to have to walk away while the cocks-of-the-rock continue the eternal contest.

The next time I went there I received a severe shock. The hide where we'd stood spellbound was a scatter of splintered wood among vigorously regrowing greenery down the slope where it had been hurled. The trees where the vivid cocks had displayed to each other and to us had been entirely swept away. From way up the slope above us a swathe had been cleared through the forest by a shattering landslide, triggered by an earthquake. The Andes are still growing and such cataclysmic events are part of their growing pains. At least nobody had been in the hide, but I could only hope that the birds were as lucky; I console myself that it's more likely that they had sensed it coming than that we would have.

Top End, Australia: monsoon forests

Cloud forests are not the only variation on the theme of rainforests.

In the tropics there are also forests, usually at low altitudes, which are subject to heavy seasonal rain followed by a long dry season – these have a very different feel from rainforests, being lower and more open, and are known as monsoon forests. In common with rainforests there are many lianas and epiphytes; in fact an alternative name, often used in Australia, is vine forest. Another, widely used elsewhere, is 'tropical dry forest'. Often some, or even most, of the trees are deciduous, due to the need to cut back on evapotranspiration to conserve water, which is a very clear difference from true rainforests.

Pittas: old passerines

These monsoon forests characterise the Top End of Australia in the Northern Territory, and further west in the Kimberley, in contrast with the always wet rainforests of east coastal Queensland and northern New South Wales. Our most recent visit was in January, far from peak tourist season. Our motive was to experience the habitat in the wet: it's less comfortable then than it is deep in the dry winter visitor season, but it's bursting with life. In this we were out of luck in that the wet was very late that year – in fact it barely came at all. On a hot still morning we were sweating our way through the monsoon forest at Darwin's East Point, ears straining, having watched the day brighten from a vantage point at the end of a walkway in the nearby mangroves. Finally there it was – a hard, ringing 'WIK WIKAWIK', coming from higher up than we would have expected. It took a while to locate the source, as the call bounced off branches and made it hard to track down the small stump-tailed black bird proclaiming itself from a branch well up in the canopy, thinned by seasonal leaf fall. Rainbow Pittas, naturally enough, are not just black, however. Through binoculars the red vent shows up, and from side-on the green-gold wings and brilliant blue wing patch glow. Pittas rarely take to the trees except to roost, spending the time hopping over the ground searching for snails, worms and other invertebrates in the litter. Pittas, some 40 species of them in Africa, southern and South-East Asia to Australia are passerines – indeed they are among the oldest groups of living passerines.

Within the passerines, nearly 90% of species belong to the Suborder Passeri, the oscines, which have the most sophisticated syrinx, the avian 'voice box'. Another Suborder, Acanthissiti, contains just two living species: the tiny and ancient New Zealand wrens. The third, to which the pitta belongs, is another old group, Suborder Tyrani: the suboscines. These form a tiny part of the bird world – except in South America. Apart from the 40 pittas, the Old World suboscines include ~20 species of Asian and African broadbills and four of Madagascan asities. But in South America suboscines rule! There are almost 1200 species of them whose ancestors prospered and evolved dramatically during the continent's long period of isolation between leaving Gondwana and crashing into North America. Astonishingly, by world standards, they outnumber the oscines, which can only muster around 800 South American species. By contrast Australia, whose biological history in most ways is eerily reminiscent of South America's, has among its nearly 350 passerines just three suboscines: all pittas. (You will have noted too

Passerines

There are some 30 Orders of living birds (the imprecision is because we are still uncovering new information as more subtle and powerful tools come to hand, and subsequently are constantly rethinking and fine-tuning our understanding of bird relationships). The most modern – that is, the most recent to have arisen – and most startlingly successful in terms of number of species is the Order Passeriformes, the passerines. This single Order provides 60% (roughly 6000 species) of all birds now on Earth. Alternatively, the other 29 older Orders between them account for only 40%, which is why we lump them somewhat disrespectfully as 'the non-passerines', defining them by what they're not. We tend to refer to the passerines as songbirds or perching birds, but, as you would expect, it's not easy to find a character that describes 6000 species. Many of them do sing, though we don't really think of most crows for instance as songsters, and some members of other Orders, such as parrots, can be quite musical. They do perch (but so do many non-passerines), though their method of doing so is instructive, with three of the four toes pointing forward and a usually long one at the back, all at the same level. (Even this characteristic is not unique, however: it is shared by birds of prey such as hawks and falcons, which are not closely related to each other or to the passerines.) However, passerines also have a special mechanism involving a leg tendon and foot muscles, so that when they land on a branch and bend their leg the toes automatically lock around the branch. Asleep, they can't fall off. There are other characteristics too, but they involve internal structures such as palate bones and wing muscles that can't be examined without cutting the bird up – rarely practical and never desirable.

Among the very first birds named by Carolus von Linnaeus, the ultimate architect of our basic taxonomic system, was the House Sparrow, which he called *Fringillus domesticus*. Two years later, the French zoologist Mathurin Brisson removed it from the Chaffinch genus and coined the genus name *Passer* for it, meaning simply 'sparrow' in Latin. This genus gave its name to the family containing it, Passeridae, and thence to the Order Passeriformes. 'Passerine' comes straight from that.

the remarkable discrepancy between the numbers of Australian and South American passerines.)

The South American suboscines comprise two subgroups, both of which we met at Angel Paz's property. The cocks-of-the-rock represent one group, which includes the South American tyrant flycatchers, cotingas and manakins, while the antpittas (no relation to 'real' pittas) front for the others, which are largely ground-dwellers such as antbirds, tapaculos and ovenbirds, as well as the arboreal woodcreepers. As a result, birding in South America represents real challenges to an outsider – it seems that so much of what we're seeing is entirely unfamiliar and it takes time to get some sort of handle on what we're looking at.

Kibale Forest, Uganda: more pittas

While thinking of pittas, I must revisit an image that stays with me, from the wonderful Kibale Forest in south-western Uganda. We set out on foot in the dark, because it is at dawn that the Green-breasted Pittas call. Our local guides, both called Gerard, were superb, winding without hesitation through a maze of faint elephant tracks in the dark, as easily as we might make our way from bedroom to bathroom at night. They carried old rifles – AK47s I am told – in case we encountered any of the elephants. It would have been a terrible thing if our presence had caused the death of one of them, and I'm not totally sure if the rifles would have been up to it anyway but, fortunately for all concerned, the elephants stayed away.

I loved watching the forest around us slowly emerge from the dark. Having reached a known pitta territory we waited in silence until Gerard heard the faint distant call, like a wooden mallet melodiously resonating on a hollow wooden pipe. As we crept closer it resolved into a double note, like the very lowest end of a xylophone, with two bars being hit almost simultaneously. Moving more quickly now we (i.e. the Gerards) located the birds on the ground – a pair and a juvenile. The birds were intrinsically gorgeous (as pittas are). A broad buffy-golden eyebrow and nape divides the black head, and a pale olive breast is sharply cut off from a crimson belly. The wings and back are a rich deep green, slashed with three of the vividest blue wing stripes. However, it is the overall context that especially stays with me: experiencing the regenerating rainforest coming to life to start another day. (We went on to spend time with wild Chimpanzees too – this really was a very special day – but that's a story for another place.)

Paluma, north Queensland: fragmented rainforest

On the eastern seaboard of Australia rainforests are scattered in suitable locales, from the steamy tropical stands of north Queensland, to subtropical forests of southern Queensland and northern New South Wales, all the way to cool temperate rainforests of southern New South Wales, Victoria and Tasmania. Certainly, they have been further fragmented since European settlement in the late 18th century, but the original isolation of these pockets of rainforest was due to climate change. For much of its isolated history, Australia was covered in rainforest: in more recent times, cooler rainforest types gradually replaced tropical-type forests in many areas. It was only 5 million years ago that rainforests disappeared from the inland, replaced by woodlands, grasslands and ultimately deserts. A new ice age began 2.6 million years ago, though we are currently in a warmer, moister inter-glacial phase within the cycle of glacials and inter-glacials that comprise an ice age. The rainforests shrank into a series of isolated fragments, mostly along the east coast. And in those fragments animals and plants became isolated from their relations, no longer able to interact and interbreed.

Riflebirds and speciation

I was thinking about that while sitting in amazed delight on the verandah of a little café in the rainforest at Paluma, in the ranges north of Townsville in the Queensland tropics. On the rail at the end of the verandah, a bird was performing, and the performance was extraordinary. The riflebirds comprise a group of four members of the bird of paradise family; they apparently arose in Australia, where three of them remain, while one has crossed to New Guinea. Victoria's Riflebird lives above sea level in rainforests from the Paluma Range north to the Atherton Tableland inland from Cairns. This one was attempting to attract the attention of a dusky female that was more interested in the bananas provided for the birds. He's not a big bird: a bit smaller than a Magpie-lark, but his presence was huge. With a characteristic harsh 'yaaahhh', which exposed his yellow mouth interior, he opened his rounded black wings and puffed up his glossy olive underside feathers, then started swinging from side to side, alternately hiding behind his flicking wings and exposing his face. It was like watching a pantomime villain swirling his cloak. An iridescent green-blue triangle on his throat and his shining crown flashed on and off behind the sooty cloak; he carefully aligned himself to make the most of the reflecting sun. I was transfixed, but though I'm sure the object of his desire peeped at him from time to time, her hormones (and his) lost out to her stomach. Had she succumbed, she would have approached him until he could enfold her in his wings before mating. In the end, he accepted that his performance wasn't impressing anyone that mattered, and he flew up into the trees.

His ancestors would have once formed a single species that ranged along the warm east coast and probably far inland. Then the rains began to fail, the winds blew cold and the rainforests retreated to sheltered refuges, each separated from the rest. There were periods of amelioration when they expanded again, but never to the point when they reformed a single vast habitat. Within the isolated fragments, the inhabitants began to change, as life does, but in directions different from those their now distant relatives were travelling. Some birds, such as the hard-flying fruit-doves, could cover the distances between the far-flung forest patches, but riflebirds are not strong fliers and have no reason to venture out into the open country. As a result, where there was just one Australian species there are now three, all isolated from each other. North of my Paluma performer, way up on Cape York Peninsula, lives the Magnificent Riflebird (though what riflebird isn't?), and far to the south, in the subtropical rainforests above and below the New South Wales–Queensland border, is the Paradise Riflebird. They all look similar but have obvious plumage differences and slightly different displays. The Magnificent Riflebird has a very different voice too: mellow rather than harsh, like a reverse wolf-whistle.

There are as many variations on this theme as there are birds in the forest, but basically this is how all species form: the essential ingredients for the recipe are time and isolation.

Cairns, north Queensland: hovering sunbird

On a rainforest edge in Cairns, north of Paluma and at sea level, a brilliant little yellow-bellied, navy blue-breasted and olive-backed bird is hovering at a red tubular flower, its long bill inserted completely in the tube to access the nectar in the gland, the 'nectary', at the bottom. I've heard it asserted that only hummingbirds can *really* hover but,

Hovering

To see a hummer hang stationary in the air at a flower or feeder, then suddenly zip in any direction, including backwards, to rematerialise and 'park' in another airy place is both thrilling and counter-intuitive. In theory at least it's relatively straightforward, involving rotating the wings in shallow figure-eights, at speeds so high that it hurts my shoulders just to think about it. Effectively they are pushing backwards at the same speed that they are pushing forwards, as well as balancing out gravity, the nett result being to hang motionless. No other bird can push forwards against the air to move backwards – they all drive only with the down/back stroke, which thrusts them forwards. For them, the forward movement of the wings (which drives a hummer backwards) is just a recovery stroke, a passive action pulled by elastic tendons. To our eyes, the hummer's wings are just a blur – even the (relatively) lumbering Giant Hummingbird manages 10 to 15 beats a second, while small ones whirr away, very audibly, up to 80 times a second.

But – and, of course, there must always be a 'but' – such a lifestyle doesn't come cheaply. The flight muscles weigh up to 30% of the bird's entire bodyweight, at least 50% more than any other bird. This is an ugly feedback loop to be caught in – more weight in turn requires even more work to carry it. At 20% of its bodyweight, a hummingbird's heart weighs, relatively, five times what ours does and beats up to 500 times a minute when resting. At full exertion, which seems to be most of the time, that doubles. At rest, the little bird breathes 300 times a minute (i.e. five times a second), some 20 times what we need. The huge energy required to drive the engine comes from nectar, which is the point of it all. To obtain the necessary 150% of its bodyweight in sugar solution, it must visit around 2000 flowers a day. For me, that would mean drinking 120 L a day, which in turn would put rather a strain on my kidneys. A hummer's kidneys are very simple to cope with the volume of work, and they constantly spray out the surplus water. Although it's almost pure water (so no need for a 'yuk' if you're standing under it), they inevitably lose some electrolytes in the process, so must eat materials such as soil and ash to compensate (e.g. Ridgely 2011; Elphick 2014, pp. 95, 408; Schuchmann and Bonan 2016).

It's a tough gig being a hummingbird, but I'm very glad that someone has to do it!

although they are indisputably the stars, in addition to the sunbirds (many of them in Africa and Asia, in addition to the Olive-backed Sunbird hovering in Cairns) I've seen small honeyeaters, notably spinebills, plus weebills, gerygones, thornbills and pardalotes, among others, hovering at flowers and foliage. (At the moment I'm ignoring 'wind hovering', which is slightly different, but we'll get to that at another time.) There may be a certain trans-Pacific superciliousness about this, but if I was used to seeing hummingbirds all the time I'd probably be a bit that way too.

Other memories of rainforests

As unpredictable as leaves side-slipping down from the canopy, here are a few more rainforest memories:

- Through a tangle of vines and dazzle of single leaves reflecting shafts of sunlight through the canopy, an enormous Southern Cassowary stands quietly in a patch of rainforest by the carpark at Mount Hypipamee on the Atherton Tableland above Cairns. There is some menace in the air, but emanating more from my knowledge that this is a potentially dangerous animal than from any threatening behaviour from the bird. The bony-looking helmet, the bare wrinkled face and neck, red and bright blue, and the swinging red wattles below the throat, all coalesce into an oddly reptilian image.
- In the green gloom of the Amazonian rainforest in Yasuní National Park in Ecuador, a large fluffy chick sits upright in the open on the end of a broken-off vertical tree branch – and is virtually invisible. It is a Great Potoo, a member of a family of seven nocturnal species related to frogmouths and nightjars. Despite our presence, the chick is very disciplined, remaining motionless with closed eyes (or maybe there's the narrowest slit there, watching us). Its pale grey downy plumage perfectly matches the lichened trunk alongside it, complete with vertical black wavy lines that mimic cracks in the bark. Further south in Amazonia, in southern Peru, an adult of the closely related Long-tailed Potoo repeats the trick, rigidly upright in full view on a broken branch stump. Mottled and streaked in browns and creams it takes sharper eyes than mine to distinguish it without guidance.
- Not far from the Potoo in Yasuní a Rufous-bellied Euphonia is singing almost into my face from only a couple of metres away. It's fair to say that from my perspective it's not an overly impressive

song (despite the genus name meaning 'euphonious'), being more reminiscent of a buzzing insect, but the bird and the situation are what grabs my riveted attention. He is shining deep blue above and orange below – and we are both 45 m above the ground (see Photo 9). However, while he is perched on a fairly flimsy branch, I am on a sturdy wooden platform attached to a magnificent old Kapok Tree. I'm not fond of heights, but the privilege of being on the same level as the canopy-dwellers is one not to be forgone.

- Thousands of kilometres to the south in Chile, in a very different Patagonian rainforest, cold and sodden, dominated by ancient *Nothofagus* trees – very similar to those in New Zealand and Tasmanian beech forests – a pair of huge Magellanic Woodpeckers is showering the mossy forest floor with chips of wood splintered from the trunk. With good reason, these birds are known as *el carpintero*: the carpenter. Beyond them, white icebergs, calved from a glacier, drift in a grey lake. The echoing of the hammering had brought us to them – magnificent black birds, the male with bright red head and wispy crest, the female with just a ring of red around the base of the bill. We watch until they fly off into the next swirl of mist and chilly rain.

References

Elphick J (2014) *The World of Birds.* CSIRO Publishing, Melbourne.

Fiske P, Rintamäki PT, Karvonen E (1998) Mating success in lekking males: a meta-analysis. *Behavioral Ecology* **9**(4), 328–338. doi:10.1093/beheco/9.4.328

Hill GE (2010) *Bird Coloration.* National Geographic, Washington DC, USA.

Marks JS, Cannings RJ, Mikkola H (2016) Typical Owls (Strigidae). In *Handbook of the Birds of the World Alive.* (Eds J del Hoyo, A Elliott, J Sargatal, DA Christie and E de Juana). Lynx Edicions, Barcelona, Spain, <http://www.hbw.com/node/52260>.

Mikkola H (2014) *Owls of the World: A Photographic Guide.* 2nd edn. Firefly Books, Buffalo NY, USA.

Ridgely R (2011) *Hummingbirds of Ecuador Field Guide.* Jocotoco Foundation, Quito, Ecuador.

Schuchmann KL, Bonan A (2016) Hummingbirds (Trochilidae). In *Handbook of the Birds of the World Alive.* (Eds J del Hoyo, A Elliott, J Sargatal, DA Christie and E de Juana). Lynx Edicions, Barcelona, Spain, <http://www.hbw.com/node/52268>.

Simpson J, Roud S (2003) *A Dictionary of English Folklore.* Oxford University Press, Oxford, UK.

Turner DA (2016) Turacos (*Musophagidae*). In *Handbook of the Birds of the World Alive.* (Eds J del Hoyo, A Elliott, J Sargatal, DA Christie and E de Juana). Lynx Edicions, Barcelona, Spain, <http://www.hbw.com/node/52257>.

Winkler H, Christie DA (2016) Woodpeckers (*Picidae*). In *Handbook of the Birds of the World Alive.* (Eds J del Hoyo, A Elliott, J Sargatal, DA Christie and E de Juana). Lynx Edicions, Barcelona, Spain, <http://www.hbw.com/node/52286>.

3

Oceans and islands

The Galápagos: oceanic islands

As I waited with my friend Juan – a top-rank Peruvian nature guide, all-round fixer and excellent travel companion – we were aware of sparrows flitting through the open airport building and heading for the *al fresco* café. Almost simultaneously, the realisation struck us – there are no sparrows on the Galápagos, and we were seeing our first 'Darwin's finches'. For a moment we stared at each other, then grinned.

Flying into Baltra airport for the first time was one of the most exciting moments of my naturalist's life. It was then an open-sided very rustic building set in a spectacularly barren black volcanic landscape by the Pacific Ocean in the centre of the fabulous Galápagos Archipelago. (Since then a flash new terminus has replaced it, very nice and admirably ecologically sound with regard to energy and water conservation, but without the same charm.) During the Second World War the US used it as a base for patrols aimed primarily at protecting the Panama Canal. When the war ended, they had no further use for it and handed it over to the Ecuadorian Government. It now doubles as a military base and civilian airport for the huge numbers of tourists who fly into this naturalists' mecca every day (see Photo 10).

'Darwin's finches'

The quotes around 'Darwin's finches' are there advisedly. The 'Darwin' appellation was not applied until well into the 20th century and, despite appearances, they are certainly not finches (or sparrows!). Charles Darwin, the unknown young naturalist who was yet to grow into the towering figure of 19th-century biology, visited the Galápagos on the *Beagle* in September 1835; he spent 5 weeks of his 5-year epic voyage there, visiting just four of the islands. His observations there famously sparked his understanding of the nature of evolution, as expounded in the scientifically pivotal book *On the Origin of Species.* Except that they did no such thing at the time! His field work on the Galápagos was, by his own later admission, not his finest hour, but being the man and scientist he was, he certainly learnt from the somewhat embarrassing experience. He did collect some finches, but he assumed that those that looked the same really were all the same – the islands from where he took them are mostly in sight of each other and he assumed the archipelago would be biologically homogenous – and he failed to keep proper notes as to the origins of the individual skins. In addition, he was more intrigued by the three subtly but clearly different mockingbird 'forms' he encountered, each restricted to its own island and obviously related to a South American species, and took his eyes well and truly off the finches.

When Darwin was in the Galápagos, he was still a conservative thinker by the standards of his own intellectual circle, and didn't accept the prospect of species changing. It was at least 18 months later, back in England, that he began to realise the significance of what he had. Even then, his first thoughts in those important directions, pondering whether species might not in time come to beget other species, were prompted by fossil mammals from the South American mainland. In those days, he thought of himself as a geologist rather than a biologist (though a naturalist then had perforce to be a competent generalist). He collected across the board wherever he went in South America: fossils, rocks, plants and animals. He had sought the advice of London Zoo taxidermists before he left to ensure his collections would be properly preserved. For the most part, his servant Syms Covington did the actual bird collecting.

Back in England in October 1836 he started the massive task of finding scientists who were capable of identifying his mountain of specimens, and available to do so. The birds were well down his list of priorities, but in early 1837 Darwin handed them to the up-and-coming John Gould, ironically a staunch creationist (a year before Gould sailed for Australia, but that's a whole other story). Within a week, Gould had come back with some startling – and, for Darwin, somewhat embarrassing – information. The apparently disparate collection of American blackbirds, finches, wrens and 'gross-beaks' all formed 'an entire new group of ground finches', with very different beaks and behaviours.

This was where Darwin was embarrassed, when it transpired that he had not seen the need to record from which particular islands each of this hotchpotch of specimens had come. Fortunately for him, he recognised the need to rectify his error, and Captain FitzRoy, FitzRoy's steward and Covington had all kept specimens and been more assiduous with recording their origins.

Darwin was a meticulous and cautious scientist, and never a healthy one, and he dreaded public exposure and even more the obloquy that he knew would come his way. It took more than 20 years for him to learn the new skills he believed he needed and to experiment, but eventually and reluctantly – nudged by the young Alfred Russel Wallace who was coming independently to the same conclusions in the East Indies – he published *On the Origin of Species* in 1859 and in important ways the world changed. There was initial resistance from the scientific and church establishments to the concept that species were not immutable, but much more resistance to the proposal relating to the mechanism – that of natural selection. Among scientists in particular, and even among the wider population, the idea of evolution had been pretty fully accepted by the time of Darwin's death in 1882; by the end of the century, it had ceased to be controversial.

The same could not be said of the natural selection concept. It left no role for a benevolently guiding deity to change species at their appointed time: everything is random, and nothing can be ordained. It took decades more for this to become a general orthodoxy.

Meantime, the finches knew nothing of this – and as far as I know, nobody has told them until this day, though they have been the subject of some of the most meticulous and exhaustive studies into the nature of evolution in a wild animal ever undertaken. Daphne Major is a volcanic crater rising from the sea within sight of Baltra – the plane you came on is likely to have flown right over it when landing. It is treeless, waterless and with no jetty or beach; researchers and all their needs must be landed from a pitching boat onto a ledge on a low cliff face. From there, all water, fuel, food, camping gear and research equipment must be carted up a steep track to the only possible camping site, which, of course, is shadeless. In 1974, the remarkable Rosemary and Peter Grant (British biologists based at Princeton), first went to Daphne Major to study the finches. For over 40 years they and their students have spent 6 months of every year on the exposed island, studying the Medium Ground Finches to the level of marking, taking detailed measurements of and DNA samples (from blood) from every single one that hatches on the island.

The environmental pressures on the birds are immense, with El Niño years of intensely cold scouring rains and La Niña times of crushing drought and food shortages. In the drought situation, birds with larger beaks do better because they can crack open bigger seeds than they usually eat, while their neighbours with the misfortune to have hatched with smaller bills simply starve. As a result, after the 1977 drought, the next generation had bills 3–4% larger than their parents' generation did. However, in the El Niño part of the cycle, the advantage swings in the other direction. The plant producing the small seeds does so prolifically, and the large-billed finches have trouble managing them; now it is an advantage to have smaller bills to harvest the seeds efficiently. The Grants' studies were among the first to follow evolution in a timeframe that even humans can comprehend. In just those 40 years, beak shape and size have changed significantly, not just in one direction, but swinging over generations on both sides of a mean. For more information in a highly digestible form, I couldn't do better than recommend Jonathon Weiner's Pulitzer Prize winning *The Beak of the Finch* (Weiner 1994).

I have circumnavigated Daphne Major – there are strong restrictions on landing there by anyone other than researchers – and wondered at many things while doing so. I have marvelled at the extraordinary dedication and resilience of the researchers and their attention to every minute detail for decade on decade in a very difficult situation; I doubt that I could even manage the landing! I've pondered the skill and experience that enables them to know every finch on the island by sight: the population fluctuates between 300 and a thousand individuals, and a new generation comes along every year. I have loved the symmetry of the story, from Darwin's eventual realisation of the profound implications of all 17 of the Galápagos 'finches' apparently arising from a single ancestor, to the Grants' work in the same place taking his ideas to a more complex plane than he'd imagined. (The number of species recognised has recently risen, following very comprehensive genome sequencing; Lamichhaney *et al.* 2015.) And I've shaken my head at the concept of the little birds surviving against all apparent probability on such a bleakly inhospitable site.

Molecular studies – including by the Grants – tell us that the Galápagos 'finches' are in fact tanagers (e.g. Sato *et al.* 2001). Tanagers, Family Thraupidae, comprise a bewildering array of mostly tropical American species. For an observer of birds, they are one of life's delights; for a taxonomist struggling to understand their relationships, well, not so much. One such authority wrote in the highly esteemed *Handbook of the Birds of the World*: 'As the taxonomic dust settles, one fact is becoming clear, namely that few avian families have been so universally misunderstood taxonomically, or have included so many taxa that have proved to be erroneously classified, as the tanagers' (Hilty and Bonan 2017). Many are splendidly coloured, but others are not. Among the less colourfully endowed are the grassquits, which are mostly plain brown little ground-dwelling seed eaters. (For the record, 'quit' is generally agreed to be a 19th-century Jamaican English term for a small bird, but no-one seems able to go deeper than that. In addition to the grassquits, there's also a Bananaquit and an Orangequit. Just saying.) The molecular studies also tell us that the closest-related living South American mainland bird to the Galápagos 'finches' is the somewhat dispiritingly named Dull-coloured Grassquit *Tiaris obscurus* from both the coast and the Andes. I have to confess that, though my records tell me I've seen this bird, and where, most atypically I cannot visualise the encounter! If one were anthropomorphic it could all be a bit sad actually.

Notwithstanding this, it seems that around 2.3 million years ago a flock of at least 30 birds immediately ancestral to both the Dull-coloured

Grassquit and the 'finches' blew from the mainland and miraculously landed on the desolate exposed lava fields that are the Galápagos (Sato *et al.* 2001) – and they would have been especially desolate back then, relatively soon after their fiery genesis. Nonetheless, the flock managed to survive and eventually thrive. Such a mishap during a storm or cyclone is probably not uncommon, but it rarely ends well.

Some descendants, notably the ground finches, retained the deep-billed seed-eating habits of their ancestors – or rather, as seems likely, lost and then regained them. Others, the warbler finches, took up an insect-gleaning lifestyle with small sharp beaks. The cactus finches, with long sharp bills, have tied their well being to the prickly pear cactuses, consuming the flowers, fruit, fleshy pulp of the stems, and insects that feed on them (see Photo 11). All of these groups have two or three species of varying sizes to further divide up the resources. Remarkably, the Woodpecker Finch has augmented its bill by using a selected twig or cactus spine to extract larvae from tree crevices for at least half of its foraging activities; moreover, the behaviour is inherited rather than learnt. The big heavy-billed Vegetarian Finch eats a range of fruit, flowers, leaves and buds.

And all this in just 2.3 million years from apparently a single flock of small ground-feeding seed eaters. Isolation and the pressures of finding ways to compete and survive in a small world have regularly led to speciation in birds finding themselves trapped on islands, but in the case of the Galápagos they had a whole archipelago uninhabited by land birds. Perhaps some of the pioneers landed on different islands originally, or over the last couple of million years some birds have moved between islands. In the case of the outlying islands in particular, it seems unlikely that this would have been due to deliberate emigration – a small land bird really doesn't want to fly out to sea unless it knows exactly where it's going – so further storm-driven mishaps were probably involved, on several different occasions.

Darwin's beach

It was an almost surreal experience to walk on Playa Espumilla (it means something like 'foamy beach') on the central Galápagos island of Santiago, knowing that 180 years previously Charles Darwin, accompanied by Surgeon Benjamin Bynoe and 'some servants', landed on this beach and walked where I was walking. Realistically, given the dynamic nature of beaches, there was more chance of me breathing a molecule of oxygen that had passed through his lungs than of my sandals walking on a grain of sand that had been in contact with his boots, but still … And some things hadn't changed. He commented on several occasions on the tameness of the birds, and when I sat in the shade a Small Ground Finch hopped right up to me. It is an amazing place, even beyond the general amazingness of islands.

Tasmania: a continental island

The Galápagos are oceanic islands: 'well, der' you might say, but this term has a specific biogeographic significance. They were never part of a continent, as Tasmania, Kangaroo Island and New Guinea are integrally part of Australia, separated for now only by higher sea levels due to cyclical melting of the polar ice caps. The cycle of isolation and re-joining of Tasmania to the mainland of Australia has repeated itself roughly every 100 000 years or so for the 2.6 million year duration of the current ice age, more formally known as the Quaternary Ice Age. Temperatures drop, more and more of the Earth's water is locked up as ice at the poles (which is by no means a 'normal' state of affairs in the longer history of our planet) and sea levels accordingly drop, by up to a hundred and more metres. The complex of causes, involving atmospheric and astronomic variables, the positions of the continents and the impact of those on ocean currents, is still debated by those far more erudite than I, and I'm happy to leave them to it. These glacial parts of the cycle within the overall ice age last for anywhere between 40 000 to 100 000 cold, dry and windy years, during which the Bass Strait becomes the Bassian Plain, mostly treeless but with some woodland on the western edges. Then, relatively quickly, much of the ice caps melt, sea levels rise, the world is again warmer and wetter and Tasmania is once more cut off by fretful seas.

Island endemism

In the tall eucalypts of the wet Tahune forest south of Hobart, along the Huon River where ancient Huon Pines arch over the hurrying waters, big Black Currawongs with piercing yellow eyes throw back their heads and emit a metallic ringing call like a trumpet blast. In an old garden in the central plateau, olive-brown Dusky Robins hunch on posts, waiting to drop onto unwary insects. On the north coast, at granite-strewn Bridport, Yellow Wattlebirds, their eponymous yellow wattles dangling beneath the base of their bills, shriek hoarsely while hunting cicadas in seaside eucalypts. And in the east, in Freycinet National Park, Tasmanian Scrubwrens approach through the tea-tree understorey to inspect us suspiciously as we picnic, and to chatter their disapproval. These are all Tasmanian endemics, which is to say they are not found anywhere else. In each of these cases, we are very familiar with their mainland close relations from which they are now separated – Pied Currawongs, Hooded Robins, Red Wattlebirds and White-browed Scrubwrens differ in plumage and voice, but in some cases not by a lot and the relationships are obvious.

Their ancestors became isolated from the rest of their kind when the ice melted and the Bassian Plain became Bass Strait again. Last time this happened was only 13 000 years ago however, and it's unlikely that all these species (and eight others) evolved to be so different in such a short time. The question is, why didn't they just meet their mainland cousins on the Bassian Plain during the many glaciations and interbreed to blend their gene pools again? I suspect that the reason lies in the conditions in Tasmania at the time, and the nature of the Plain. During the most recent glaciation, for instance, the snowline was 1000 m lower than it is today – both in Tasmania and on the mainland – and a 6000 km^2 ice field, up to 70 m thick, covered Tasmania's central highlands. Birds were not ranging across the landscape, but huddling into remnants of forest in sheltered lowland situations. The mostly treeless plain was not attractive to forest-dwelling species. Technically, birds could have crossed in both directions, and some probably did (after all there are many species which do occur on both sides of the stormy strait), but for many more species exploration was not as much a priority as was simple survival in such harsh conditions.

Tasmanian Scrubtit: an older endemic

Hobart, Tasmania's capital, is a beautifully situated small city on the sheltered mouth of the Derwent River, snuggling up into the forested slopes of imposing Mount Wellington. In sheltered lower gullies, big old Soft Tree Ferns (often called Man Ferns in Tasmania) crowd for space, and a delightful network of walking tracks traverses the slopes. Above the ferns, a small rusty brown bird, like a Tasmanian Scrubwren but with pale grey cheeks, has flown up from a tree fern and is working up a eucalypt trunk, probing its curved bill into crevices for insects. This is another endemic, the Scrubtit, but, unlike the others, it has no obvious mainland counterpart and indeed is the only member of its genus. It is possible that there was a mainland member of the genus that died out, but it is more likely that the Scrubtit evolved from a mainland ancestor many glacial cycles ago and has been a Tasmanian resident for longer than the other endemics. Intuitively, we might expect that ancestor to have been a scrubwren, but a recent paper surprisingly suggests, based on molecular studies, that it probably evolved from a whiteface (*Aphalocephala* spp.): three species of dry country birds no longer found in Tasmania (Gardner *et al.* 2010).

The 'turbo chook': a cautionary tale

Another Tasmanian endemic bird that any visitor will surely come across is the slightly manic-looking hyperactive Tasmanian Nativehen, flocks of which will run from roadsides and ovals on sturdy legs, jinking and weaving to shelter in reed beds or in water. They have forgone the use of their wings for flight, relying instead on those strong grey legs to avoid predators, their wings swinging out to balance them as they change direction. Big birds, half a metre high, they are reputed to run at up to 50 km/h and this strategy obviously works, because they still roam in big flocks in places, despite serious persecution in the past for being 'agricultural pests' – they were only afforded legislative protection in 2007. Tasmanians refer to them fondly as turbo chooks: 'chook' being a universal Australian word for domestic hens, from an old English word 'chuck' or 'chucky', presumably from their clucking.

(I sometimes fear, however, that it might be part of a uniquely Australian idiom vanishing into a homogenised television English.)

This chook has as its nearest relation the nomadic inland Australian Black-tailed Nativehen (also fondly known as 'Barcoo bantams'): a smaller bird, not flightless but mostly found running across the ground with an erect tail like the turbo chook's. They will appear in huge numbers in good conditions and vanish again overnight. Clearly, they form another in the now familiar pattern of species pairs created by the filling of Bass Strait (see Photo 12).

Well, no, actually. 'Tasmanian' Nativehens were formerly widespread on the south-east mainland until the last glaciation (i.e. from 20 000 to 12 000 years ago), when it seems that the increased aridity killed them off – though why that didn't also eliminate them from Tasmania remains an unanswered question (Boles 2005). However, that too seems to have been a premature conclusion, because, more recently, mainland turbo chook remains from just 4700 years ago have been unearthed (BirdLife International 2016), so they were surviving until then, albeit in apparently much lower numbers, when the Dingo apparently arrived from Asia with seafaring Macassan traders. It is surely no coincidence that after this date the chooks only survived in Tasmania, where the Dingo didn't appear. It seems that the day-active chooks could manage to survive alongside nocturnal predators, such as the Tasmanian Devil and Thylacine, but not the day-hunting Dingo.

All of which means that the nice neat pattern suggested by other Tasmanian endemic birds isn't a template for all its endemics: did the turbo chooks evolve on the mainland (and, if so, why did they lose their flight powers there?) and later run to Tasmania, or did they evolve in Tasmania and cross to the mainland during an earlier glacial period? For now, we simply don't have the fossil evidence to answer this question, but it remains a useful reminder to us not to be too complacent about one-size-fits-all explanations for island endemics – or, indeed, for anything else!

One possible excuse for such complacency in this situation (though we would still be wrong!) is that flightlessness is a characteristic of island bird faunas, especially among the rail Family to which the Tasmanian Nativehen belongs, so it is reasonable to assume they arose in Tasmania.

Having fortuitously arrived at an island, like the grassquit ancestors of the Galápagos 'finches', rather than perishing at sea, there is considerable evolutionary pressure to make the most of your luck and stay there. Not that I'm suggesting any self-determination among the birds, but if I were to fall off a cliff and miraculously be saved by a protruding cliff-dwelling tree (you know how it works from watching just about any action movie), I wouldn't be viewing this as a sign that it would be OK to jump off again. In the birds' case, we'd expect that any genetic tendency to head out to sea again is likely to represent a dead-end line, because the chance of finding land a second time is pretty small.

Lord Howe Island: a volcanic speck

Lord Howe Island comprises the eroded remains of a volcanic crater: a speck of crescent-shaped rock, some 10 km long and 7 million years old, in the Pacific Ocean 600 km east of New South Wales. Strangely and atypically, those great seafarers, the Polynesians, seem not to have discovered the island, so that when Europeans settled there in 1834 to found a supply station for the whaling industry it was still pretty much a pristine wilderness. (Although the loss of its innocence had begun earlier than that, when Lieutenant Henry Lidgbird Ball went ashore in February 1788 to claim it for England, and begun the plunder by taking a load of turtles and 'tame' birds back to Sydney. Later visitors deliberately introduced pigs and goats onto the island to succour any stranded sailors.)

Island extinctions

Among the first birds encountered were two species of flightless rail, like the Tasmanian Nativehen. One of these, the Lord Howe Island Woodhen, is still with us (though it very nearly wasn't – more on that anon). The other was a strange and beautiful big white bird with large red bill, the Lord Howe Swamphen or White Gallinule. This species scarcely survived to see the 1834 settlement; passing sailors ate it to extinction. There are just two skins of it, both in Europe, to represent the entire evolution and violent passing of a whole species – one is in Vienna, the other in Liverpool. This sad story was just the beginning of the Lord Howe bird disaster, however. The White-throated Pigeon (a subspecies of a south-west Pacific species) was abundant and was slaughtered in vast numbers. It was so tame that it was killed by sticks, and it had gone by the middle of the 19th century. The Lord Howe Parakeet made the mistake of eating the newcomers' crops, and paid with extinction by 1869 (Hutton 1991). (There used to be some comfort in the belief that the parakeet belonged to the same species as the Norfolk Parakeet, which just clings to life on Norfolk Island, but opinion has changed on that; its passing represented the quenching of another entire species.)

Nonetheless, there was still an abundance of small bush birds, famously tame and enriching the forests, gardens and even houses of the settlement. Or they did until 1918, when a disaster even greater than the earlier ones arrived without anyone even noticing. In that year, the island trader *Makambo* ran aground and its cargo of Black Rats swarmed ashore. Astonishingly, within a decade, five endemic passerines, three or four of which were full species, all common and some abundant, had gone for ever. No-one will ever again enjoy or even see the Vinous-tinted Thrush (now regarded as a subspecies of Island Thrush), the Lord Howe Fantail, Lord Howe Gerygone, Robust White-eye or Tasman Starling. (There used to be a subspecies of the starling on Norfolk Island, but we dispensed with that population as well, so the species has gone for ever.) Even then we had not quite finished our fatal meddling. In a belated attempt to curb the rats (remember the *Old Lady Who Swallowed a Fly*?), Australian Masked Owls were introduced from Tasmania, along with some Barn Owls and, inexplicably, mainland Southern Boobooks. They didn't get rid of the rats, but by the 1950s the endemic Lord Howe Boobook (a subspecies of Southern Boobook) had gone, probably out-competed by the two smaller invaders, and quite likely eaten by the much bigger Masked Owls.

I tell this sad story not because it is an atypical one, but because it isn't. Islands have been witness to tragic bird stories for centuries now.

The Lord Howe Island Woodhen: almost not there and back again

Perhaps 200 bird species have become extinct since the 16th century – the exact number has to be an approximation, because some of them were not formally recorded or collected, though descriptions and even illustrations exist. This could lead to both over- and under-estimates, but any such errors probably balance out. Moreover, some apparently extinct species could, with luck, still exist in remote areas. On the other hand, some other species seem to be sliding inexorably towards extinction. At least three species (the Socorro Dove from an island off Mexico, the Guam Rail and the Alagoas Curassow from Brazil) exist only in captivity; others, including Spix's Macaw, are teetering on that same brink. *And 90–95% of these extinct species were island birds* (nearly a third of them Hawaiian). One in six of all these recently extinct birds were flightless island rail species, like the Lord Howe Swamphen and Woodhen (e.g. Szabo *et al.* 2012).

The Lord Howe Island Woodhen is an engaging olive-brown bird the size of a small chook. It potters about, poking its long bill into ground litter and rotting logs to extract worms, grubs and arthropods. Its response to an unfamiliar object is likely to be to wander straight up to it, no matter if the object happens to be a cat, dog or hungry sailor or settler. This is one of the problems of evolving on an island where there were probably no significant predators for most of your history there – you don't have any appropriate defences or reactions. Cats, dogs, pigs and settlers depleted the numbers until, by the late 1970s, there were fewer than 40 woodhens left, restricted to the remote peaks of Mount Lidgbird and Gower. When a survey in 1980 found just 15 individuals, drastic action was required, and thankfully was taken. Birds were taken into captivity, where fortunately they were quite willing to breed. At the same time, urgent research into the several possible causes of the continuing decline identified feral pigs as the key threat, and their elimination was commenced. The overall program has been a rare and impressive success story: woodhen numbers are up to 250 birds, which, with a required territory of 3 ha per breeding pair, is probably the most the island can carry (e.g. Frith 2013). I have met them carelessly crossing the road from the airport into the adjacent heath, forcing us to stop our hired bicycles, rustling under the bushes outside the island bottle shop (I can't now recall why we were there) and exploring the lawns of our guest house.

It is no coincidence that such a high proportion of recent bird extinctions has been among the island rails, and that so many of them are flightless. The woodhen is in the same genus as the Buff-banded Rail – it is a matter of some debate as to just what that genus is, but it is generally agreed that, whether it's *Gallirallus* or *Hypotaenidia*, they're both in it. This is relevant because the Buff-banded Rail is an amazing wanderer across watery wildness. It seems as though there's something in its genes or psyche that drives it to take off and head out to sea, or perhaps it is just congenitally careless about taking to the air in storms. This is despite the fact that we don't often see it flying – it tends to skulk around thickets and wetlands. But fly it most certainly does: there are (or were) at least 22 subspecies on islands of the south-west Pacific, from the tropics north of the Philippines to the sub-Antarctic (though the Macquarie Island subspecies is no longer with us) and east to Fiji, Samoa and Tonga. This is an astonishing feat of distribution in a bird not noted as a strong flyer, and it is not at all surprising that it or a closely related ancestor reached Lord Howe Island at some point – indeed, in recent decades, newly arrived Buff-banded Rails have also started to build a substantial population on the island (Hutton 1991).

Island breeding colonies: from Fernandina to Cape Town to Victoria to Chile

I had the good fortune of meeting the Galápagos Flightless Cormorant at Punta Espinoza: the only permitted landing place on Fernandina, the western-most and youngest island of the archipelago. The birds, numbering only a few hundreds, live only here and on adjacent Isabela. Volcano Cumbre is still very active, coughing up lava every few years, and the land surface is dominated by black volcanic rock, with smooth swirls in parts and jagged edges in others. Clusters of Lava Cactus, one of the first colonisers of newly cooled land in the Galápagos, sprout seemingly (and actually) from the rocks. Tight masses of short thick somewhat phallic stems form dense colonies: the packed spines are bright yellow when young, darkening with age. Hundreds of Marine Iguanas, dark as the lava, sprawl in piles to warm in the sun between sessions of browsing on algae in the cold sea currents.

Why give up your wings?

There are, it seems to me, a couple of reasons why a genetic adaptation, such as flight, might be subsequently lost, despite the obvious advantages it confers. One eventuates if the species begins to 'experiment' with a new lifestyle that interferes with flying, making it first difficult, and eventually impossible, as the new skill becomes more important than the old one. One example is that of penguins, which became so adept at 'flying' underwater that their wings became ultimately suitable only for flight in that medium, rather than in the air. Another is the ratites – the giant ancient Gondwanans including ostriches, emus, and so on – whose 'experimentation' was with progressively greater size for greater safety, until the universal cut-off point of 15 kg was reached, beyond which true powered flight becomes mechanically impossible.

The second possible reason why the power of flight may be forgone is represented by a conundrum: what if the environmental pressures that led to its evolution disappeared? If a bird found itself in a situation where it had no enemies, then the cost of maintaining its flight might eventually prove to be not worthwhile. This would be especially true if there was nowhere to fly to. The answer to this somewhat Gollumesque riddle is the situation facing a group of probably unwilling colonists on a remote island. As we have previously observed, flight on a small island may in fact be a hazard, with the risk of being again blown out to sea. Many different unrelated groups of birds, in addition to the rails, have flightless members on oceanic islands (e.g. the New Zealand Kakapo is a parrot and the Dodo of Mauritius and Rodrigues Solitaire were pigeons).

The Galápagos has a flightless cormorant and there are two flightless teal species on New Zealand islands, and a flightless steamer duck on the Falklands (as well as a couple on the mainland of far southern South America). To a weakly flying waterbird, large alpine lakes might be considered islands in a hostile sea of heathland, and there are two flightless grebes on high-altitude Andean lakes in Peru, including on the huge Lake Titicaca (see Photo 13). Other species, such as the Kagu of New Caledonia and the mesites of Madagascar, have been isolated for so long that their relationships are obscure, though the Kagu seems to be closest to the equally enigmatic Sunbittern of South America. The mesites, which are almost flightless and seem to be evolving in that direction, may be distant pigeon relatives, but the necessary DNA work appears still to be done.

Tragedy after tragedy has ensued when an enemy – inevitably people and their animal associates – has suddenly appeared in the habitat of a bird that has forgone its best protection: flight. Victims have included the moas, the elephant-birds and the Dodo, as well as the lovely snowy Lord Howe Swamphen and the numerous other flightless rails.

These cormorants are the world's largest: a metre long, dusky brown with a long tail and bright blue eyes. They nest on a sandy-rocky spit, with the substantial seaweed-pile nests scattered just above the high tide mark so that each is out of reach of its neighbours. When the non-brooding partner comes ashore from fishing, the birds reinforce their partnership with extravagant head gestures and gnarring displays. Like other cormorants, they lack waterproofing oils in their feathers – oil is lighter than water, and would make their underwater fishing too difficult – so must hang their wings out to dry, when their puny nature becomes evident. It is obvious these wings could never carry them through the air (see Photo 14).

This colony numbers only a dozen or so pairs, but seabird nesting colonies can be huge. Lambert's Bay is an industrialised little fishing town on the west coast of South Africa north of Cape Town, but is notable for hosting a massive colony of Cape Gannets. The access to the well-managed and well-interpreted colony is very easy to miss: a little back street leading past a factory and warehouses to a carpark from where we walk across a causeway to a small island – imaginatively named Bird Island – where a hide (with a window downstairs and open on the first floor) looks straight into a solid mass of black-winged white birds with yellow-buff necks and caps. They press in (some 30 000 of them when I was there, according to the chalk board in the hide) as densely as they can fit. But seabird colonies rarely comprise just one species. At Lambert's Bay, behind the gannets and among the Cape Fur Seals, are breeding Cape and White-breasted Cormorants and Kelp Gulls, and Crowned and Bank Cormorants and Hartlaub's Gulls also breed on derelict boats in the adjacent harbour.

In south-western Victoria, the pretty little town of Port Fairy has more historic buildings per square metre than almost anywhere in Australia. It also has an even greater treasure, which, like Lambert's Bay's treasure, is on a little island joined to the mainland by a causeway. Among the grass clumps on Griffiths Island are up to 100 000 burrows excavated by Short-tailed Shearwaters (also widely known as Muttonbirds for their contribution to a long-standing industry of harvesting the oil-rich chicks on islands off Tasmania). One parent broods in the burrow, while the other fishes at sea – each day at dusk large numbers of the foragers come home to change shifts. If you wander down to the viewing area around sunset you can look out to sea and see dim hints of movement on the horizon, like distant swarms of insects or perhaps just a trick of the fading light and straining eyes. When it's almost too dark to see, and the more impatient or uninformed visitors have wandered off for a meal, suddenly the air is full of large dark bodies plummeting to earth and the burrows, being greeted by a great chorus of breathy squeals from underground. It's a strangely moving experience.

Across the Atlantic from Lambert's Bay and on the far side of South America, Puñihuil is a small fishing community on the wild Pacific coast of the island of Chiloé in southern Chile. Assisted by the Alfaguara Project (a marine conservation organisation focusing on the Blue Whales that breed in these waters), the local fishermen take visitors out in open boats to sail around the three little islands in the mouth of the bay. It is a spectacular experience, with the skill of the fishermen holding the boat seemingly perilously close to the rocks shrugging off the waves, and the cacophony and wild reek of the nesting colonies merging with the thump and hiss of pounding water. This is the only place where cold-water Magellanic Penguins and tropical Humboldt Penguins breed together. They are joined by four cormorant species, the stars of which for me are the Red-legged Cormorants, which I reckon to be the world's most beautiful – ash grey with bright red legs and red-based yellow bills, a big white neck patch and silver-spotted black wings. Unlike either the scattered Fernandina Flightless Cormorants or the crowded Cape Gannets of Lambert's Bay, Red-leggeds' nests are just out of pecking distance of each other, which is more the norm in such colonies. Additionally, the islets support breeding Kelp Gulls, Kelp Geese and Fuegian (Flightless) Steamer Ducks.

The basic reason for such colonies is simple necessity – there are finite suitable breeding sites, which for safety are on islands or cliffs, so they must lump in with other individuals and species. Overall, some 13% of bird species nest in colonies, but over 90% of seabirds do so. There are specific advantages too, though. For instance, colonies can provide group surveillance and defence against predators. By having huge numbers of eggs available at once, they can make it impossible for predators to take them all; in this case, the best sites will be those in the centre of the colony. But – and, of course, there must always be a 'but' – every strategy has a disadvantage. If there were a perfect strategy everything would use it! Problems associated with colonies include the fact that they are so easy for a predator to locate, shortages of nest sites and building materials and potentially of food to sustain a large number of mouths. Birds such as gulls threaten their neighbours with cannibalism, and disease and parasites spread very rapidly. The risk of wasting effort (often an entire day's work at sea) in mistakenly feeding the neighbours' offspring is perhaps the greatest of all. All colony-dwellers must be able to recognise their own eggs and young among the hordes, by location or voice.

Chile's Chonos Archipelago: prolific seabirds

The Coastal Cordillera of Chile is a less lofty range of mountains running parallel to the Andes, but nearer to the coast. Between the two ranges is the huge and immensely fertile Central Valley from where pour Chile's excellent wines. The cordillera runs into the Pacific just south of Puerto Montt – Chiloé Island comprises its peaks protruding from the sea. South of Chiloé the cordillera continues as the complex scatter of islands that make up the wind-hammered Chonos Archipelago. Vessels, from fishing boats to cargo ships to luxury cruise liners, ply the channel between the Archipelago and the snowy Andes peaks and glaciers that line the shore. Unless the clouds are lying on the sea so that islands and volcanoes are behind a densely opaque curtain, the views are superb, and the channel waters are alive with birds. Huge flocks of Sooty Shearwaters float in rafts on the sea, pattering across the surface with a very audible rush of water and soaring just above the waves, sliding across the bows of the ship. Pink-footed Shearwaters are more likely to be single, but are in pretty much constant view. Less common petrels (the Family that includes shearwaters) such as chequered Cape Petrels and palest grey stocky Southern Fulmars appear as luck determines. Enormous louring Southern Giant Petrels, big dark solid

White-chinned and Westland Petrels, as well as some tiny ones in the vastness, delicate Wilson's Storm Petrels and chunky Magellanic Diving Petrels, all come and go for as long as you can brave the wind in your eyes. And, from time to time, the vast and utterly majestic form of a Black-browed Albatross soars effortlessly across the grey heaving world.

Wind soaring

But when we turn into one of the fjords that indent the coastline, something peculiar happens – all the birds disappear. Not for them the sudden calm of the sheltered inlets: they crave and need the wind to ride upon. Indeed, both petrels and albatrosses seem to have arisen some 40 to 50 million years ago in the vast southern oceans, where the winds never rest. If the wind falls below ~20 km/h, an albatross must wait on the water. Although shearwaters and other petrels in particular can fly hard and fast when they need to (especially during migration), for the most part they let the wind do the work for them. To watch a shearwater or albatross delicately slice the length of a wave top with one wingtip, then swerve through 180°, swing up high from the water, and suddenly be gliding low and fast over the surface again a couple of hundred of metres away, without once flapping, is to doubt one's eyes.

The secret is in those amazing wings. All of the birds we have just watched, or at least all the bigger ones, have very long slender wings ('high aspect ratio' in aerodynamic terms), with each square centimetre of wing carrying only a very small weight ('low wing loading'). This is the perfect shape, theoretically and practically, for slow flying without crashing, and for soaring. Soaring is sometimes dismissed as just a sideshow to true powered flight, but in reality it is very complex behaviour indeed, and can be seen as the epitome of flight, achieved in all the long history of wings by very few, mostly large, birds and pterosaurs. Indeed, some ancient giant pterosaurs and birds could soar but could no longer truly fly. Soaring involves using the energy in the surrounding air to stay aloft, while expending very little of your own energy. Some land birds use warm air columns ('thermals') such as may be found rising over sand dunes or even buildings to achieve great altitudes from which they may glide, slowly descending to the next, or simply soar in circles watching for prey. They may also use wind deflected upwards by a dune or mountain range to the same effect.

The Chonos Archipelago seabirds, however, are using a different energy source: the layers of wind over the sea blowing at different speeds at different heights. Simplistically, the layers closest to the sea surface are moving more slowly due to friction; waves increase the effect. The bird uses its wings and the relatively light wind close to the surface to lift itself off the water (often aided by some running on the surface), then climbs slowly through the layers into the wind to gain altitude. By swinging through 180° and plunging down through the layers again, it can gain energy as it goes. An albatross or shearwater can keep this up for days and thousands of kilometres, flying across the wind. And, as ever, there's a 'but'.

The world's sole tropical albatross, the Waved Albatross, breeds only in the Galápagos, and only on the island of Española (well, actually, a few dozen also breed on the little island of La Plata, closer to Ecuador – but that's all). The cold waters are rich in food and the trade winds that blow from March to December keep the albatross aloft. In January and February, however, the winds drop and the albatrosses must move to cold Humboldt Current water off Peru and Ecuador. The problem comes when they want to take off from the ground at the nesting colony – and this is why they only nest on Española, a dot on the north-eastern edge of the Galápagos. Their wings (as with any high aspect ratio wings) are simply too long to flap without bashing into the ground, so they must run from the nest to the cliff edge, and hurl themselves off into the wind. Only on Española is there enough flat land to support a colony, and near enough to a cliff edge that faces into the prevailing wind. All of life seems to be about trade-offs!

Other memories of oceans and islands

Like droplets of spume drifting ashore from a wave destroying itself on granite rocks, a few arbitrary images of the sea and its islands:

- Standing on the deck of a small cruise boat, with less than 20 of us aboard, we are sailing between islands in the Galápagos.

Overhead a small flotilla of black Magnificent Frigatebirds floats effortlessly: males with a purple sheen to the neck feathers and flaccid red throat pouch, females with white throat feathers. They are huge, with a span of well over 2 m and long forked tails, slender wings angled into an 'M'. For as long as they accompany us, benefiting from the updraft of air from our passage, not one flaps its wings. Not once. A female descends so that she is sailing alongside us, one alert eye level with mine and watching me.

- Another experience of watching soaring seabirds at eye level was while walking the Malabar Cliffs on the north coast of Lord Howe Island. Here the stars are the beautiful snowy white Red-tailed Tropicbirds, which breed on the cliffs below us in huge concentrations and drift in glorious parade along the cliff face, pairs performing their mesmerising courtship dance. In these flights, one bird drifts along while the other 'stands up' on the air, pushing so that it moves backwards, while each points its tail towards the other (see Photo 15).
- Pisagua, on the desert coast of far northern Chile, is a town of ghosts. Once a bustling port with 10 000 people and a concert hall that drew world-class performers, now mostly empty with a couple of hundred people relying on the fishing. Moreover, a series of 20th-century brutal military dictatorships used Pisagua as a prison camp for non-criminals – gay men in the 1920s, communists in the 1950s and various leftist opponents of the regime in the 1970s and '80s. Many are buried outside the town. The cold Humboldt Current, bringing nutrients up from the ocean depths and swathing the coast in mist, is one of the richest parts of the planet, creating the world's most productive marine ecosystem. It produces some 20% of the global human fish catch, and bird and other animal life abounds. In the 19th century, Pisagua supported a huge industry mining the seabird guano, and the birds remain. On the day we walked along the cliffs to Punta Pichalo, we watched thousands of Guanay and Red-legged Cormorants, Peruvian Pelicans and Brown Boobies streaming around the point and racing to intercept a vast school of small

fish moving out to sea. The sheer numbers of birds and their desperate energy to be in on the feast were riveting.

- In a small open boat bobbing in the swell just off the extraordinary sheer cliffs of Balls Pyramid, south of Lord Howe Island, a group watches entranced as Flesh-footed Shearwaters and White-bellied Storm Petrels come in to feed on berley (bait) thrown over the stern. The storm petrels, 'Mother Carey's chickens' of sailor lore, are tiny and almost never seen from shore. They are pattering on the water surface in their characteristic way, holding their wings up to catch the wind and using their feet in the water to anchor themselves in place for a few seconds at a time (Withers 1979). It is enthralling to watch such behaviour that is normally inaccessible to land-based creatures.

References

BirdLife International (2016) *Tribonyx mortierii*. The IUCN Red List of Threatened Species 2016, <http://dx.doi.org/10.2305/IUCN.UK.2016-3.RLTS.T22692900A93373971.en>.

Boles WE (2005) A new flightless gallinule (Aves: Rallidae: *Gallinula*) from the Oligo-Miocene of Riversleigh, northwestern Queensland, Australia. *Records of the Australian Museum* **57**(2), 179–190. doi:10.3853/j.0067-1975.57.2005.1441

Frith C (2013) *The Woodhen: A Flightless Island Bird Defying Extinction*. CSIRO Publishing, Melbourne.

Gardner JL, Trueman JW, Ebert D, Joseph L, Magrath RD (2010) Phylogeny and evolution of the Meliphagoidea, the largest radiation of Australasian songbirds. *Molecular Phylogenetics and Evolution* **55**(3), 1087–1102. doi:10.1016/j.ympev.2010.02.005

Hilty S, Bonan A (2017) Tanagers (*Thraupidae*). In *Handbook of the Birds of the World Alive*. (Eds J del Hoyo, A Elliott, J Sargatal, DA Christie and E de Juana). Lynx Edicions, Barcelona, Spain, <http://www.hbw.com/node/52380>.

Hutton I (1991) *Birds of Lord Howe Island Past and Present*. Ian Hutton, Coffs Harbour, New South Wales.

Lamichhaney S, Berglund J, Sallmän Almén M, Maqboo K, Grabherr M, Martinez-Barrio A, *et al.* (2015) Evolution of Darwin's finches and their beaks revealed by genome sequencing. *Nature* **518**(7539), 371–375. doi:10.1038/nature14181

Sato A, Tichy H, O'hUigin C, Grant PR, Grant BR, Klein J (2001) On the origin of Darwin's finches. *Molecular Biology and Evolution* **18**(3), 299–311. doi:10.1093/oxfordjournals.molbev.a003806

Szabo JK, Khwaja N, Garnett ST, Butchart SHM (2012) Global patterns and drivers of avian extinctions at the species and subspecies level. *PLoS One* **7**(10), e47080. doi:10.1371/journal.pone.0047080

Weiner J (1994) *The Beak of the Finch: A Story of Evolution in Our Time.* Random House, London, UK.

Withers PC (1979) Aerodynamics and hydrodynamics of the 'hovering' flight of Wilson's Storm Petrel. *The Journal of Experimental Biology* **80**, 83–91.

4
Mountains

Torres del Paine NP: Andean Condors

Not much could have distracted my enthralled gaze from the stunning panorama in front of me, but the huge black bird drifting between me and the view did it easily. I had wanted to see an Andean Condor since it featured in a book on animals that I'd absorbed and treasured as a small boy. An annoyingly precocious small boy, I must add (at least when it came to animals – in all else I was hopelessly shy and introverted). It is part of family lore that when my Glaswegian-born grandfather invited me to 'look at the wee birdie' in the book I politely corrected him with 'actually Grandpa, it's a Female Condor' (I thought that 'female' was part of its name).

The lookout at Lago Nordenskjöld in Torres del Paine National Park in Chilean Patagonia is one of my favourites in the world. It never ceases to astonish and delight me. You come on it after driving for some time through the park: a vast wind-pruned landscape where herds of Guanacos graze and skitter. Pull over, get out into the inevitable wind and gaze across the pale green ruffled waters of the lake to Los Cuernos – the Horns. (The milky blue-green water that characterises this part of the world is due to glacier-ground 'rock flour' in suspension.) Although the wind just ruffles the water on a good day, I've also seen the surface of the lake carried as a cloud of spray above the waves by wind that can knock the unwary off their feet. The Horns are truly spectacular: sheer pillars of granitic rock that rear up from the far lake shore. The lowest slopes are cloaked in tough Antarctic Beech (*Nothofagus*) forest but above that is bare cliff, a broad pale band above the dark base and capped with another hard dark layer (of basalt), with snow lying in gullies. Then gradually the scale sinks in. The peaks we're looking at are more than 2 km above us: the apparent 'snow faces' at the end of the gullies are actually glacier walls more than 50 m high. The clarity of the air seriously messes with one's perceptions.

This is a vast landscape and it somehow seems appropriate that a bird that lays (disputed) claim to being the world's largest flying bird finds its last stronghold here. It's also a bird that can tell us lots of stories.

Let's start with the size, and it really is enormous! The male has a wingspan of up to 3.2 m, and weighs up to 15 kg; the female is somewhat smaller. It's generally accepted that 15 kg is about the upper limit that a flying bird can be. True, there were pterosaurs and even birds heavier than that – the mighty *Argentavis magnificens*, which cast its shadow over the Argentinian plains 6 million years ago, weighed perhaps 70 kg – but these could neither have taken off nor stayed aloft by flapping (Ksepka 2014). They would have relied on either launching from a cliff or running downhill, and using air currents to stay up – they soared rather than flew.

Featherless heads

The mighty condors roost and breed on cliff faces and simply fall into space, after which they rarely need to flap, circling in the ever-busy air currents over the mountains. Not far from Nordenskjöld lookout, I once came across a remarkable sight: up to a dozen condors on the ground, feeding on the fresh carcase of a young Guanaco, or chulengo, perhaps killed by a Puma the previous night. One huge male exerted his dominance, lowering his great head to the ground, ruffling his white-collared neck and raising his wings to show his white secondary feathers and wing-coverts as a white panel against the black, and flushing his bare face a deeper red. Usually only one adult male is present at such a gathering and actual conflict is rare. Both sexes have featherless heads and necks, but the female's is black and the male's is red with a fleshy comb from eyes to bill, with wattles on his neck.

Many birds have bare heads, for the same general reason – if you're going to dip your head into something wet and sticky you don't really want feathers to be all gummed up. This applies to various ibis and stork species that probe into mud, some friarbirds (large honeyeaters) that stick their heads into large nectar-rich flowers and both New

How big can a bird be?

A brief basic physics lesson, and trust me that it will be basic (Physics 1 was the only university subject I failed). As a bird gets 'longer', it also gets heavier, but at a much higher rate. A bird twice the length and wingspan of another has four times its surface area (i.e. it increases by the square of the length); this is a plus, because bigger wings mean more push. *But* it also has eight times its weight (which increases by the cube of the length). In other words, wing loading (weight per area of wing, which determines how hard it must work to fly) increases with size – despite larger wings, a larger bird must work twice as hard to take off as the smaller one half its size. A bird three times the size must work three times as hard, and so on. Take-off is aided by jumping and, in a larger bird, legs are as important as wings at the moment of take-off. But, jumping is much easier for a lighter animal: compare the relative heights that a flea, a frog, a wallaby, a horse and an elephant can jump. A large bird of prey such as a condor or a Wedge-tailed Eagle, especially if full of carrion, may have to run along the ground to take off; a swan or pelican or albatross must run along water.

Taking off is the really hard bit, but it's by no means the end of the problem. A large bird must also fly faster to stay up, so it must have relatively larger wings and flight muscles (as well as just proportionally larger), which in turn add extra weight. It's a vicious circle that seems to vanish down its own plug hole at around 15 kg. You must then rely on soaring (once you've managed to get off the ground), but the heavier you are the faster you fall, so you must find stronger and stronger air currents to lean on. There are reports of rare individuals of very large birds such as Kori Bustards and Mute Swans that weigh 20 kg or more, but it's not clear whether those atypical birds can still get off the ground.

Another intriguing suggestion has been made about the upper limits to size, and it is independent of weight. All birds must regularly replace feathers, which quickly become worn and inefficient; this process is called moulting (more on this on pages 189–90). In 2009, Sievert Rohwer and colleagues at the University of Washington published a paper showing (very) mathematically that although feathers of larger birds must be longer than those of smaller ones – roughly double the length for a 10 times increase in mass – feather growth rates don't increase accordingly (Rohwer *et al.* 2009). A point is reached at which the replacement feather can't grow quickly enough to serve the bird's purpose. For instance, a 10 kg bird needs 8 weeks to replace a single flight feather. With around 10 of the big primary flight feathers (though some larger birds have 12), it would clearly be impossible to moult them only one or two at a time. The options are to moult several at once, with the resultant extended disadvantage of inefficient flight for a long time, or to lose them all at once, meaning at least 2 months of being unable to fly, or to extend the moult over 2 or even 3 years. Many albatrosses take the latter option, but, even for them, the effort of breeding and moulting is too much in one year, and so they avoid breeding in consecutive years to focus on the moult (Langston and Rohwer 1996).

World and Old World vultures, which get much of their intake from the interior of large carcases. But (have I mentioned there's always a 'but'?), it's hard to explain the spectacularly bald heads and necks of wild turkeys, some guineafowl and the West African picathartes or rockfowl, for instance, as adaptations to feeding. Most of these do have colourful head and neck skin, which plays a role in display, but it's hard to say if this is sufficient to actually dispense with feathers, which can also be brightly coloured. One study on the Griffon Vulture of north Africa, the Middle East and central Asia concluded that its featherless head was related to temperature control (Ward *et al.* 2008), but it's hard to imagine how this could be beneficial to the Andean Condor in the very chilly Andes and Patagonia. However, it *could* be if the function was later adapted to other purposes.

A condor wishing to assert authority – like the big male at the chulengo – can, as noted, flush bright red to make his point. It was assumed that such displays were based on red pigments in the skin that evolved for the purpose, but Negro *et al.* (2006) showed otherwise for a diverse array of birds – in at least 20 Families across 12 Orders, most of which are large dark-coloured tropical species including cassowaries, vultures, caracaras, bustards and parrots. In many cases at least, the flushing is due to highly vascularised skin close to the surface, which originally developed to assist in heat loss; nearby skin under the feathers has no such characteristic. As species such as the condor came to use this feature also to signal information such as status and fitness, perhaps the value of this secondary use came to outweigh the disadvantage of an uninsulated head as some of the birds headed south.

It is widely suggested too that exposing the skin to ultraviolet light and dryness helps control bacteria from the condor's restaurants, but while highly plausible it's hard to find evidence for this.

The awesome alula: making flight possible

As another condor landed hoping to get a share, she spread her huge wings and tail and feet to act as brakes, and raised a tuft of feathers on the angle of each wing. It would be easy to miss if I wasn't watching her through my binoculars. This little tuft, the alula (AL-yoo-la), is arguably one of the most significant developments in the history of bird evolution.

'Alula bird'

Cuenca is an ancient walled city some 150 km from Madrid, with World Heritage classification. However, to a birder (not to mention a bird!) it has an even greater significance. From nearby, in the mid-1990s, a remarkable fossil was reported, of a superficially unremarkable bird about the size of a scrubwren. This bird, *Eoalulavis*, lived some 115 million years ago. The detail of the fossilisation makes me shake my head with wonderment at the sheer, beautiful unlikelihood of it. We know what it ate, because its stomach was still full of shrimps. And we know that it truly flew, as truly as any modern bird does, because its superb fossil feather impressions included an alula! (*Sanz et al. 1996*). For the record, *Eoalulavis* means 'dawn alula bird'.

But what *is* an alula and how does it work? As I have already confessed, I have no credentials at all in physics, and in any case even the professional aerodynamics people can't seem to agree on the details of why a wing (be it of bird or aeroplane) actually works – at least as far as I can understand them. So, this explanation will of necessity again be simple. As the leading edge of a wing tilts further up (or if you'd rather, the 'angle of attack' increases), two key things happen. First, the lift increases (partly due to decreasing pressure on the top of the wing, as it forms a 'pressure shadow' behind the raised leading edge, so that the wing is forced up): this is essential if the bird is to accelerate and climb. However, the drag also increases as the area facing the wind does: this pulls the bird back, so it must flap harder to produce extra thrust to compensate. A wing held horizontal to the ground attracts least drag, but it will also not allow climbing so is of limited use.

It is important to reduce turbulence over the leading edge of the wing, in part because turbulence breaks up the relative vacuum above the wing and increases drag. However, as the angle of attack increases, so does turbulence. This is manageable under most normal flying conditions, but there is one aspect of every flight where the angle of attack must be raised to almost 90°, with potentially disastrous results. This is at the moment of landing, where wings, tail and feet are spread fully out to deliberately increase drag, in order to decrease speed to stalling – just as I saw in the condor coming down hoping for a snack of chulengo. In such circumstances, turbulence over the wing would be catastrophic, simply dropping the bird out of the air with all the grace of their wingless reptilian forebears. But now the marvellous alula is spread, like a mini-wing stuck up from the real one (a perhaps unfortunate alternative name is 'bastard wing' – I find alula much more euphonious). Air flows through the slot formed and over the wing surface. For reasons well beyond me – and, I would tentatively and respectfully suggest, apparently beyond those much better credentialed than I, though it seems they're getting closer – this air flow is nice and smooth and the bird, rather than stalling or crashing, touches lightly down (e.g. Lee 2015). (See Photo 30.)

The alula is controlled by the first of the three remaining much-reduced 'finger' bones in a bird's wing (but technically this is number two; one and five have disappeared entirely). An aeroplane achieves the same thing by means of wing flaps and slots; the superb and sorely missed pteranodons did so by lowering a flap of skin across the front of the wing using a single small tiltable bone; bats use a network of tiny hairs across their wing surface to inform them of minute changes in air flow and prompt them to subtly alter the wing shape (Sterbing-D'Angelo *et al.* 2011). None could successfully fly without an alula-equivalent.

Archaeopteryx apparently had not evolved an alula, so presumably hit the ground running, or crashing. It almost certainly couldn't have landed on a branch from other than a short hop, or pounced onto prey, because it couldn't have slowed its descent enough to do so. Somewhere in the brief few million years between *Archaeopteryx* and *Eoalulavis* the first alula appeared. No living birds descend from *Eoalulavis,* so either the alula evolved more than once or it arose very early in the story of birds, in a hitherto unrecognised species that was the common ancestor of every bird we see today. Either is possible, and either way it emphasises the critical importance of the alula in the long history of bird flight.

Can birds smell?

Unlike the much more widespread Turkey Vulture, the condors have no very useful sense of smell – they find their meals by sight, which is not so hard on the open country of the mountains. On the other hand, the Turkey Vulture might be better at finding carcases in general, but

Birds with a sense of smell

We now know that many birds really do have a good sense of smell, and some of them a very acute sense indeed (e.g. Balthazart and Taziaut 2009; Rajchard 2008; Averett 2014; Corfield *et al.* 2015). Kiwis have exceptional abilities to sniff out food, especially earthworms, in night-time forest soils: a kiwi's olfactory lobe accounts for a full third of its brain according to some accounts. The Turkey Vulture and its relatives can detect ethyl mercaptan, which is released by decaying flesh. (This skill has been used by pipeline engineers: ethyl mercaptan is added to natural gas to enable humans to identify a gas leak, so vultures are experts at finding a leak in a natural gas pipeline!)

Another group of birds with a remarkable sense of smell is the petrels, which we now know can find dimethyl sulphide (or DMS, a chemical released by crushed krill, as happens when birds or other animals are already feeding on them) at sea from tens of kilometres away. They follow the scent, zig-zagging upwind, to the source. Moreover, we also know that the DMS is released by krill not just randomly across the open ocean, but especially where underwater features such as mountain ranges are close to the surface, which is where plankton tends to accumulate. With this information, the petrels not only can follow their beaks to find food, but develop a map of the oceans (e.g. Nevitt and Bonadonna 2005; Nevitt 2008). A downside of this skill is that various plastics, which are dumped at seas in vast quantities, react with algae to release DMS, which encourages seabirds to eat them and starve to death (Savoca *et al.* 2016).

Various members of this Family, which nest in burrows and return to them in the dark, recognise their own home by its distinctive aroma. And a final twist to this story: in 2012, it was shown that European Storm-Petrels can reliably distinguish relatives from unrelated birds by scent, and consistently preferred the unrelated ones when it came to choosing a mate. This was the first time this had been demonstrated, but given the importance of avoiding inbreeding as far as is possible, it probably won't be the last (Bonadonna and Sanz-Aguilar 2012).

that alone isn't enough – they need the massive condors to open a large carcase before the smaller birds can access it.

It is ironic that it is the Turkey Vultures, along with the closely related two yellow-headed vultures, which have the power of smell, because it was the experimental observations of them by US bird artist and would-be ornithologist John Audubon in the 1820s that virtually closed the door on any further consideration of the concept of bird olfaction for a century and a half. Having observed the vultures attack a meat-less sewn-up deer carcase, then circle above but reject a hidden rotten pig body, he deduced that they couldn't smell anything. What he could – and indeed should – have otherwise deduced was that the vultures did indeed find the deer by sight, but also found the covered pig by scent, whereupon they decided it was just too ripe for them! It turns out that they prefer their carrion to be no more than 4 days old. The opinion among some of the modern olfactory biologists, whose quest to have their field taken seriously was long in vain, is that Audubon should have stuck to his day job. In the forests the three scent-detecting vultures come into their own. They can find a corpse on the ground beneath the canopy as easily as they can in the open; the big King Vulture and abundant Black Vultures must keep an eye on them and follow them down to the prize.

The rear-guard action fought, even if only in a passive way, by people who think that Audubon was basically right, but have been forced to concede that kiwis, vultures and petrels are exceptions, has been surprising. These bird groups are not exceptions to a general lack of sense of smell among birds, though they are especially good at sniffing out the world. One of the problems is that it has proven more difficult in birds than in mammals to predict their olfactory expertise from the shape of their brain, and it's taken a while to realise that, for a bird, this is quite normal. Back in 1993, a paper was published showing that, despite a lack of evidence in the shape of their olfactory lobe, a randomly selected (by mist netting) group of five small passerines had a capacity to recognise and respond to scents, which was as acute as that of rats and rabbits (acknowledged experts in the game) (Clark *et al.* 1993).

Birds may not have cute whiffly noses, but don't ever let anyone tell you that they can't smell…

K-selected breeders: 'have one kid, make sure it survives'

Andean Condors nest high in their rocky fastnesses, of necessity. Getting off the ground is a huge effort for a bird that is at the 15 kg limit, beyond which flying becomes very challenging (and after they've gorged more kilograms at a carcase they may be flightless until it digests). Accordingly, they roost and nest on cliff ledges, which are abundant in the peaks and outcrops of Torres del Paine, and along the vast spine of the Andes. In ecological terms, they are K-selected breeders, which means that they rely on the world being a fairly stable place in which life is long and death rates are low (see page 124 for more on the history of the term). In this ideal world, they can afford to have very few young and put a lot of effort into ensuring that they survive. In a human-free world, condors may live for up to 50 years, not breeding until they are 5 years old, and produce just one egg every 2 years. It is laid on an exposed rock ledge between 3000 and 5000 m above sea level; it takes 8 weeks to hatch, then the chick needs another 6 months to fledge, after which it remains dependent for another 2 years.

There's not much room for change or error in such a plan (with only one egg every 2 years), but sadly the condor's world *has* changed, and much for the worse. Torres del Paine is one of their strongholds, and they're doing fine in the southern Chilean Andes and north to northern Argentina, but beyond that the story becomes grim. Populations have fallen badly in Bolivia, Peru and Ecuador, and to critical levels in Colombia and Venezuela. The problem comes when farmers see condors on a dead cow and assume they killed it – a story that played out for decades of tragic and pointless slaughter of Wedge-tailed Eagles in Australia. Ironically, condors are still honoured on the national coats of arms of Chile, Bolivia, Ecuador and Colombia.

Such breeding systems evolve over millions of years: it is not possible for such a long-lived, slow-breeding species to adapt to a different strategy in human timescales. All the big K-breeders, among which the plight of albatrosses is especially well known, are potentially at risk; the condors are just one glaring example among the many.

Colca Canyon: high-altitude birding

As I said earlier, condors can provide the entry to so many stories, and the condors of Colca Canyon are no exception.

Colca is an extraordinary place, some 3700 km north of the snowy peaks of Torres del Paine, well into the tropics in the arid highlands of the eastern slopes of the southern Peruvian Andes. Coming from the desert coast, the road climbs to the daunting Abra (meaning a pass) Patapampa, 4900 m above sea level. To those of us unused to such altitudes, the few steps up to the lookout present a challenge to already shocked lungs looking for enough oxygen (needless to say, the local women selling crafts by the roadside don't seem to notice it!). The landscape is shockingly stark to our eyes – devoid of green, spurs of jagged volcanic rock protruding from the sand and stretching away to the snowy volcanoes emitting drifts of smoke on the horizon and towering 6000 and more metres into the sky. Not many birds are evident here, though I didn't have much spare energy to go looking either! From here the road descends to the east, but not to altitudes that our lungs immediately recognise as 'normal'.

It passes above the ancient village of Chivay through a landscape that has been subject to intensive agriculture for over 1000 years, well before the ascendancy of the Incas (though they tend to get credit for all Andean culture and technology). Terracing and irrigation supported the growing of corn, potatoes, quinoa and beans, as well as grazing flocks of Llamas and Alpacas. The pre-Incan terraces are still prominent, and are still cultivated. At these lower altitudes (though still 4000 m above sea level), vegetation is dominated by a diversity of cacti and bromeliads, including lichen-like air plants and aloe-resembling puyas with big flower spikes. Andean Flickers (big handsome tawny ground-dwelling woodpeckers with brown and white barred back and wings, black moustache and bright red nape) hop across the ground and mount old stone walls. Dull-coloured ovenbirds such as Slender-billed Miners (which gain their name from nesting in burrows) search the track-sides for insect snacks. Bare-faced and Black-winged Ground Doves rest in cactuses among spines long enough to completely transfix them.

Then, the canyon. It is famed for being among the world's deepest, at its maximum 3300 m straight down from the lip to the Colca River, twice the depth of the Grand Canyon, as the Peruvian publicity material is understandably fond of telling us. At the lookout where we and many others pull up early in the morning at Cruz del Condor (Condor's Cross) Lookout, the drop to the river is a mere 1.2 sheer kilometres, but it's impressive enough. Around the extended lookout grow the bushes bearing the long bright red tubes of Peru's national flower, Qantuta (*Cantua buxifolia*), regularly visited by the plain-coloured but imposing Giant Hummingbird, which is positively lumbering when compared with its petite relatives.

Soaring: ultimate flight

However, the cars and coaches that constantly pull into the car park are not here for any of these delights, because at Colca Canyon roosts one of the last major concentrations of Andean Condors in Peru. They spend the night on rock ledges far down the precipitous walls and, as the air begins to warm in the morning, condors and air currents wake up and both begin to climb up from the depths to the sunshine. It's never certain that the condors will rise in front of the viewing platform – there is after all a lot of canyon on both sides of here – but that morning we were in luck. It started with the hint of movement in the shadows far below, which resolved into the unmistakable bulky shape of a condor circling in the faintly stirring air, searching for the energy in the breeze to carry it up towards us. Over the next half hour or so, we were treated to the thrilling experience of 17 condors at close range, up to half a dozen at a time spiralling up to the day, soaring in big circles to gain height, and finally dispersing along the canyon and across the drab brown countryside (see Photo 16).

The silhouette of a soaring condor cannot be mistaken for anything else: massive, with a short wide tail and very broad wings tipped with seven immensely long separate 'fingers' comprised of the elongate primary feathers. Condors are consummate soarers, as this column of rising birds attests, not deigning to flex their wings into anything resembling flapping. Indeed, while most of these birds are youngsters and adult females, an adult male at close to 15 kg couldn't survive by self-powered flying alone, as we have observed previously. But when we first looked at soaring, in albatrosses and petrels soaring over the oceans using the updrafts from waves and layers of moving air getting faster with height, we admired the very different shape of their wings. Those very long slender wings with low wing loading (weight borne per square centimetre of wing) are the most aerodynamically perfect shape for soaring, a realisation used by designers of gliding aircraft. So why do the condors not have such wings?

The answer is in the *length* of the 'perfect' soaring wing. We saw how the Waved Albatrosses at the Española breeding colony could only take to the air by running to the cliff top and launching into the prevailing wind. If they tried to launch into the air by flapping, their very long wings would bash against the ground and it simply couldn't work. Most land birds can't rely on a handy cliff to jump off, so a compromise is necessary – and that compromise is the accurately, if inelegantly, named 'soaring wings with slots' as demonstrated by the soaring condors. They are not the perfect soaring wing, though they can have lower wing loading than the albatross wing model, but they still endow the bird with good soaring skills while also enabling take-off from the ground or from freshwater surfaces. A condor, especially gorged on dead chulengo or cow, will probably need to run downhill while flapping to get airborne, but it can be done. A pelican will need to run across the water surface to build up momentum, but its wingtips won't be dragging in the water.

Back at Colca Canyon, the last condors finally dispersed to do a day's scavenging work and we left to do a walk along the canyon rim. Meantime, tour buses were still disgorging clients. I wonder if their guides hadn't mentioned condors to them, or if they were going

'Soaring wings with slots'

The condor's basic wing shape (long and broad with those long 'fingers') is repeated in a range of quite unrelated large soaring birds. Pelicans, storks, ibis, cranes, eagles and Old World vultures have all quite independently evolved this 'next best' compromise wing shape for soaring. The broad wings and tail 'sit' on the air, absorbing its rising energy, while the slotted tips break up the turbulence around the ends of the wing, keeping the air smooth in much the way that the alula does. (For a detailed discussion of just how they do it, taking account of changing air turbulence, see Reddy *et al.* 2016.)

In central Queensland I have watched a dozen big Australian Pelicans and a huge lanky Black-necked Stork rising in a wide spiral above the Thomson River at the delightfully named small town of Muttaburra. All had essentially white wings – except for the tips of the feathers, both secondaries and primaries. Blacks, greys and browns are provided by a group of pigments called melanins, and it so happens that melanins also confer a significant resistance to wear, meaning that black wing tips are a very useful soaring accoutrement indeed.

to attribute the lack of condors to just bad luck as opposed to their own tardiness.

The Drakensbergs, Lesotho: snow in Africa

I hadn't really associated Africa with snow but, of course, that was just my ignorance. I've since seen it from Cape Town airport in the jagged ranges to the east, and in the wonderful Drakensberg Mountains in south-eastern South Africa. Here they also include the high-elevation little nation of Lesotho (formerly Basutoland), entirely enclosed by South Africa. The Drakensbergs differ from the Andes in a couple of significant ways, including being much less extensive and, at a maximum 3500 m above sea level, not as high. (Still, compared with Australia ...) The Andes are still growing, forced ever upwards as the Nazca Plate pushes under the South American Plate along the Pacific coast of the continent. Earthquakes, landslides and volcanic eruptions represent Andean growing pains. The Drakensbergs ('Dragon Mountains' in Afrikaans, though no-one seems quite sure why) finished their growing long ago, and are slowly eroding away. The genesis of their final stage came when Gondwana began to fracture some 140 million years ago and molten rock squeezed up through the resulting cracks. This volcanic material cooled to form the hard cap to the older sedimentary rocks, leaving the Drakensbergs rearing above the surrounding lowlands to the east.

The road from Kwa-Zulu Natal to Lesotho follows the infamous Sani Pass: 9 km of gravel road (though I gather that sealing is planned

Tolkien and the Drakensbergs

It is widely known and reported that J.R.R. Tolkien, whose father worked in a bank in nearby Bloemfontein, found his inspiration for the grim mountains of Mordor in the Drakensbergs. Isn't it sad though when facts leap out of the shrubbery to scare a perfectly good story to death? (I guess that's why there is a strong tendency in some places these days to eschew facts in favour of an interesting yarn, no matter how improbable and even disprovable it might be.) In this case, the tedious fact is that young Tolkien left South Africa at age 3 and never returned. He much later apparently reported that his only memory of the country was of a huge spider: perhaps Shelob, rather than Mordor, was South Africa's contribution to the Rings trilogy (SouthAfrica.net 2016).

for the future) climbing 1000 m up a long series of hairpin bends with precipitous drops below and black and white parapets, rock and snow, looming from the clouds above. There can be ice, snow and mud on the road at any time of year, and towards the top some pinches are at close to a one in three gradient. It is said that if you look down you can see in the gorges below the remains of vehicles that didn't meet the criteria for success: I tried to look up at all times! This road is for 4WD vehicles only, though the Lesotho border authorities at the top are pretty relaxed about who they allow to descend. Given that the pass is in South Africa, I guess they reckon that it's not really their problem. At one point on the road we came across an old truck broken down on a nasty switchback corner – apparently it had been there for 5 days – and we had to get out to walk cautiously past it, while vehicles ground around it in lowest gear, perilously close to oblivion.

At the top of the pass in Lesotho, Africa's highest hotel stands at 2860 m above sea level on the snowy plateau, snow piled on the roof and icicles hanging low from the eaves. Sloggett's Ice Rats (a hamster-like delight), Cape Sparrows, Speckled Pigeons and Drakensberg Siskins forage in snow outside. Just before that huddles the Lesotho border post: a small grim square stone building in a bleak white landscape. But what flew over was special indeed.

Bearded Vulture: digesting bones

The Lammergeier, or Bearded Vulture, is huge, with a wingspan of nearly 3 m and white-speckled black wings, tail and back, lovely orange neck and feathery trousers and a distinctive black face and beard below the bill. Despite the name, it is not especially closely related to the true Old World vultures. It has a huge, but sparse, distribution in the mountains of Europe, the Middle East and central Asia, and north-west and eastern Africa. I had never seen one, and would have felt I'd lived a very slightly poorer life had I never done so.

But what is really remarkable about the Lammergeier is its diet, up to 90% of which comprises bones! Both 'how?' and 'why?' jostle for attention but I'll start with the how before the why leads us onto distant paths far from Lesotho's chilly plateau. Perhaps we think of bone as being something akin to rock when we're considering food items but, even apart from the nutritious marrow, it does have food value – if you can extract it. In fact, a study on captive Lammergeiers at Tel Aviv University (part of a breeding program for reintroduction to the wild) showed that those birds preferred bones to meat when offered the choice (Houston and Copsey 1994)! Moreover, they liked old bones more than fresh ones, presumably because they are easier to break up, and because they become lighter as they dry out, so are easier to carry away. We also know they prefer larger ones to small ones (Margalida 2008). The birds were extremely effective at accessing the organic component of the bones (sheep ribs, by the way, if you'd like to contemplate that as a diet, without the meat attached!). The Lammergeier's task is to extract the organic component, which comprises mostly short collagen fibres (the protein that forms connective tissue), from the matrix of calcium phosphate apatite, which is of limited use to the bird, and it does so with near total efficiency. Remember, too, that there is the bonus of the fatty marrow at the end of this, but the bone itself supplies most of the bird's nutrition.

The Lammergeier doesn't perform the task by keeping the bone in the gut for a long time – the canal is not significantly longer than that of meat-eating raptors – and it can digest a rib bone in just 24 h. Rather, it has a dense layer of acid-secreting cells in the oesophagus and stomach, producing a highly acidic environment, akin to battery acid. As the bone softens (and remember the Lammergeier can't chew it up like a hyaena, so must swallow it whole, lengthwise), muscular contractions break it up. In this way, they can deal with bones up to 250 mm long and 35 mm in diameter (Elphick 2014). Lammergeiers have reportedly even been seen flying with the end of a bone protruding from the mouth, waiting for the other end to digest so they can swallow it all!

It is also well documented that, when faced with bones too big to get down whole, it will carry them into the air and drop them from heights of up to 150 m onto favoured and repeatedly used rock platforms. Bones of 4 kg (half to two-thirds of the bird's weight) have been recorded as being lifted and shattered in this way, and multiple attempts may be required.

Amazing, but why bother? Vultures, despite a common prejudice, actually don't like very putrefied meat, though their harsh internal environment destroys many of the harmful bacteria ingested with the meal, and they can tolerate others to a surprising degree. There comes a point, though, when the carcase is no longer an acceptable food source. Moreover, they have serious competition in the form of other bird and mammal scavengers, bacteria, fungi and insect larvae, so after a while all that is left is ... bones. These last for a very long time, so a Lammergeier living in the high mountains, where the density of large mammals and thus their carcases is low, can greatly extend its food supply by being able to use a single carcase for weeks or months until all the skeleton is gone. It's a bit like having a deep freeze to keep fresh your single bulk purchase until you've eaten it all.

Torres del Paine again: New World vultures

Unexpected origins

Inevitably we've wandered far indeed from the condors demolishing the chulengo carcase in the icy, windy mountainscapes of Torres del Paine, so let's go back there to pick up another story they have for us. The question to introduce this story is a pretty basic one – 'what are they?'. Well, as we've noted, they are the largest of the New World vultures, but that's just begging the question, which is 'what are the New World vultures?'. It's now well established that they are not just Old World vultures that found their way to the Americas. In fact it seems they are not very closely related at all, despite appearances and behaviour. The vultures of Africa and Asia belong to the same Family of birds as eagles and hawks, the Accipitridae; the American vultures belong not only to a different Family (Cathartidae) but, according to many taxonomists, to an entirely different Order of birds. Their resemblances are due to parallel lifestyles, and animals (including birds) tend to look the way they do because that makes them *good* at what they do.

We can readily imagine Old World vultures deriving from a hunting bird such as a Wedge-tailed Eagle or Whistling Kite (to use Australian examples, though there are equivalents elsewhere too), which heavily supplement their diet of prey with carrion. But what was

Other almost impossible foods

The Lammergeier may be the most extreme example of a bird specialising in a resource that is unavailable to others (no other vertebrate, even the Spotted Hyaena, relies so heavily on bones), but it is not unique. Others have cracked the secrets of using potential foods that are physiologically unavailable to most animals. Ants and eucalypt leaves, for instance, are both appalling foods, both chemically and physically, but their saving grace is in being an almost infinite resource, so, if you can solve the problems inherent in digesting them, you are guaranteed a food supply. Woodpeckers and Australian treecreepers are two groups that prey extensively on adult ants but, although many insect-eaters snap up the odd ant, very few make an effort to do so consistently. There are many South American passerines, including antbirds, antpittas, antthrushes and antwrens, that follow the army ant columns (in some cases exclusively) but they're after the insects terrified into motion by the ants, not the ants themselves. On the other hand, no bird eats eucalypt leaves; indeed very few birds eat leaves at all, but that's a topic for elsewhere (see page 139).

The 8000 species of sawflies (wasp relatives) are found across much of the world. In Australia, there are fewer than 180, of which many, not unexpectedly, have learnt to eat eucalypt leaves. They deal with the toxins by separating them from the edible components and, rather than wasting them, add insult to the plant's injury by storing them as caustic yellow oils in the foregut to regurgitate in times of their own need (hence the name 'spitfire'). Not many birds will risk this chemical warfare, but one seems unfazed by it. The Gang-gang is Australia's smallest cockatoo: ashy grey with pale mottles and bars and a wispy crest comprising separated spiky feathers like starfish arms. Moreover, the males have a clownishly bright red face, head and crest, and their voices creak like corks coming out of bottles as they fly overhead or sit quietly feeding. They are an eternal delight and I often have the pleasure of their company at home, because Canberra is the only city where they can be found throughout the urban area, including the city centre.

A particularly fond memory of them, however, is of a male sitting in a leathery-leafed Snow Gum high in the Snowy Mountains of Kosciuszko National Park in the Australian Alps of south-eastern Australia. Alongside him was a cluster of steel-grey tubular spitfires, frantically tapping their heads on the leaf in warning to their colleagues and dribbling yellow fire. He was quite unfazed as he perched on one foot, picking up the larvae one at a time in the claws of his other, delicately extracting the gut sac with the tip of his beak and dropping it to the ground before thoughtfully eating the rest like a connoisseur of eucalypt lollies.

Spitfires closely resemble caterpillars: the unrelated larvae of moths and butterflies. Most of these have limited defences beyond camouflage, but quite a few have evolved irritating bristly hairs or long thin ones whose tips can pierce skin and be shed, or even contain toxins. These are pretty effective against most predators, but not all. Many cuckoos and cuckoo-shrikes (this combination is a curious coincidence, since they are entirely unrelated, in separate Orders) specialise in hairy caterpillars. Cuckoo-shrikes have learnt to rub the hairs off on branches before eating them, though cuckoos are also able to eat large quantities of the hairs without ill effect, regurgitating them in pellets. As with the spitfires and the Gang-gang, these birds aren't so much eating these awkward unpleasant meals *despite* the fact that others don't, but *because* they don't. Like the ants, hairy caterpillars and spitfires represent a major untapped food supply for which there is little competition; if you can stomach them, you'll be a lot less likely to go hungry.

A few other birds have also opted for such a diet that is downright hazardous and quite unavailable to almost anything else, giving them a free run at it. There are four 'honey buzzards' in the genus *Pernis* (one breeding in Europe and three in both temperate and tropical Asia). They are not true buzzards and don't eat honey, so best not to enquire into 'what's in a name?' in this case. Their chosen dietary niche is one of the most forbidding imaginable – the nests of wasps, hornets and bees, which they raid as a major part of their diet for larvae, pupae and the waxy comb itself. They dig into the ground for buried nests and take nests from trees. To answer the obvious question, it seems that small stiff dense feathers, which are especially scaly on the face, protect the skin from stings (Birkhead 1974). Moreover, Sievwright and Higuchi (2016) have suggested that a 'unique filamentous substance' on feathers, especially around the eyes, could be evidence of chemical defence, but this has not yet been established.

It is regularly reported that the honey buzzards eat the waxy comb, but it is not clear whether this is simply in order to extract the contents, or if they do digest the wax itself. If so, this would be a most unusual situation: wax certainly contains nutrient value, but it is hard to extract. In fact, only the honeyguides (a small Family of 16 species in the same Order as barbets and toucans, from Africa and Asia) are known to have mastered the trick as a group. Although earlier studies suggested that bacterial colonies were employed for the purpose, more recently unusual enzymes including lipases have been found in some species (e.g. the Lesser Honeyguide; Downs and van Dyk 2002) and there now seems to be doubt about the bacterial theory. It has been suggested that some seabirds can also do so (to cope with waxes in crustaceans), as well as some passerines that eat wax-coated berries.

Incidentally, the honeyguides don't generally adopt the full-frontal assault tactics of the honey buzzards: although their skin is unusually thick, it is not impervious to stings, and birds have been found dead with numerous bee stings. They tend to attack the hive early in the morning before the bees are active, or find abandoned hives, which apparently are more common in hot climates than temperate ones. They will also take 'leftovers' from hives that have been attacked by other animals, especially humans – the behaviour of Greater Honeyguides in southern Africa in leading people to a hive is well documented.

More specialists: fitting the bill

So far, these examples haven't involved the development of a specialised bill for eating unconventional food, which is a bit surprising given that the bill is the primary food-gathering organ of a bird. Such bills certainly exist, however. One unusual bill form, which has evolved on several occasions for different functions, involves a long slender upper mandible that significantly overlaps the lower one.

The Red-capped Parrot is endemic to the south-west of Western Australia, and is the only member of its genus (perhaps closest to the rosellas). Although it also eats a range of seeds and fruits, it specialises in the tiny seeds of Marri (*Eucalyptus calophylla*), which are held in a large, hefty cup-shaped woody seed case up to 50 mm long. The parrot nips off the entire fruit and holds it in one foot, anchors it further with the broad-tipped lower jaw, and delicately extracts the seeds from the cup with the thin upper mandible (Forshaw 2002) (see Photo 17).

The Long-billed Corella is a medium-sized white cockatoo with a red-orange face and breast band, and a massive bill with an elongated upper mandible. It lives in open woodlands and grasslands (mostly now dramatically altered by farming and grazing) of inland south-eastern Australia. It uses its bill not for delicate extraction, but for serious excavation into the soil: when a flock has been working a paddock it has very obviously been cultivated! The Murnong, or Yam Daisy (*Microseris lanceolate*), once covered vast areas of Australian temperate grasslands and was a key food item for the Long-bills. Accounts from southern New South Wales and northern Victoria (i.e. Long-billed Corella heartland) tell of swathes of Murnong flowers turning the plains golden to the horizon. Their small sweet tubers were harvested by Aboriginal people, eaten raw or roasted to a delicious treacly consistency. European settlers learnt the trick from them. There are stories of wagon wheels turning up thousands of Murnong tubers from the soft soil, leaving them to rot on the surface. Then the sheep came, eating the plants and learning to push into the soil to eat the tubers as well. The plough finished the job. Initially, Long-billed Corella numbers crashed, until they learnt to eat the wheat that replaced the Murnong. This was, of course, a capital offence and populations declined again as large numbers were shot. Now they survive largely on exotic lily-related tubers, for which their specialised bill is pre-adapted.

In South America, and to a minor extent in central America, the Caribbean and Florida, two raptor species also have similar bills – and, almost without exception, raptors are carnivores, with no interest in seeds or tubers. In this case, the food for which their bill evolved presents a similar problem to the Marri fruit – large, smooth and hard with a small opening. The big apple snails (*Pomacea* spp.) are found widely in South America and provide an excellent food source to anything that can breach their defences: a challenge that is beyond the capacities of most birds. Snail and Slender-billed Kites, however, use their long sharp upper mandible to cut the columellar muscle that holds the snail's body to the shell, after which the meaty delicacy just falls out.

Among the very few other birds that have learnt to breach the defences of apple snails are the two species of openbill storks (one African, one Asian): in their cases the snails are *Pila* spp. These storks have a most peculiar bill indeed, in which the lower mandible bows outwards towards the tip, and is twisted slightly sideways, creating a distinct gap of several millimetres when the beak is closed. There has been a lot of debate as to how they use this structure to extract the snails, not least because the process is both rapid and mostly occurs under water. It is now agreed, however, that they do not break the shell, or use the gap to carry snails away. Rather, they use a concentration of stalk-like pads near the tip of the straight upper mandible to hold the snail in place against the ground or mud, then use the angled lower mandible to push past the operculum (the hard plate that protects the shell opening) and, like the American kites, to cut the columellar muscle. Perhaps even more astonishingly, the Asian Openbill produces a narcotic chemical in its saliva, which flows down the bill and relaxes the operculum muscle, facilitating access (Elliott 2016).

It's important to remember that all these specialists can, and do, also eat 'normal' food (i.e. as defined by their less specialised close relatives) if circumstances warrant it, but use their 'super powers' to escape competition from those lacking them.

the ancestor of the condors and other American vultures? There is still not consensus (not an unusual situation with taxonomists, who deal with vast stretches of time, often incomplete evidence and evolving investigative tools), but the numbers seem to be increasingly behind an old idea: that their closest relatives are storks (they aren't storks, but share a common ancestor). It's not hard to envisage a bird with the carrion-eating lifestyle of a modern Marabou Stork giving rise to a dynasty of carrion specialists (Elliott 2016).

Even the apparently logical notion that the New World vultures evolved to fill an empty niche caused by the absence of 'real' vultures is at odds with reality. Until just 10 000 years ago, there were Old World vultures in America, and there were New World vultures in the Old World, though not apparently for the last 20 million years.

However, the concept of birds evolving into something quite different to fill an empty niche is not at all fanciful. Take the Southern Crested Caracaras waiting patiently for their turn beyond the scrum of condors on the carcase. This is quite a substantial bird, with a wingspan of up to 1.3 m and weighing over 1.5 kg, but it is dwarfed by the condors. It has short feathered trousers, its back and flight feathers are brown-black, while most of the rest of the body is barred with fine white markings on a dark background. A white throat and cheeks are topped by a black cap and blue bill with a broad red base. The bill alone leaves no doubt that it is a bird of prey: either vulture or mainstream raptor. BLAA! (That was the quiz show buzzer announcing our humiliation and loss of points for this incorrect assumption.)

This solid predator, which spends a lot of time walking across the ground in search of food as well as scrounging carrion, is a falcon. Not even 'evolved from a falcon ancestor', but a fully paid-up member of the Family Falconidae. Nonetheless, it eats almost anything, from mammals to birds to reptiles to frogs to crabs to worms to insects to carrion – not at all a falcon's diet. So what niche does it fill that was empty enough to tempt a falcon from the skies to the ground? The answer seems to be that of crows. When I first encountered a flock of the little brown Chimango Caracaras working over the ground in the Patagonian heaths, I was irresistibly reminded of Little Ravens in the high country of south-eastern Australia foraging in just the same way. When South America collided with North America just 3 million or so years ago there were no corvids (i.e. members of the Family Corvidae, such as crows, ravens, jays, magpies and nutcrackers) in the southern continent. While other vertebrates, such as North American blackbirds and camels, spread throughout South America, it seems that the crows foundered in the vast rainforests of the north and, even today, the only corvids in South America are a few species of rainforest jays. This left the vast plains of the south with an opportunity apparently too good to refuse, and falcons filled it. There are still mainstream falcons in South America too: it's just that some of their ancestors chose another path.

Flying high: how do they do it?

Andean Condors are found to at least 5000 m above sea level (about the level of Patapampa Pass, where I struggled to walk a few steps), but that's nowhere near the limits of altitude for birds. There are reports from planes of migrating birds and big raptors at double that altitude, but at 10 km above the ground each breath contains less than a third of the oxygen that is available at sea level. Most of us could not survive just being there, let alone undertake prolonged strenuous exercise. In an experiment some 40 years ago, mice and sparrows from sea level were put into a chamber with an atmosphere equivalent to 6100 m (i.e. 20 000 feet, hence the arbitrary metric figure). The mice were prostrate, as would we be, but the sparrows appeared unaffected (Tucker 1968). How do they do it?

Birds have had to develop a rather different principle of respiration from mammals, because their lungs are smaller and unable to expand very much due to the constraints of the great flight muscles and associated supporting bones. Though a bird's lungs are relatively smaller than a mammal's, the air capillaries are finer and more numerous, giving a greater surface area. In fact, a duck's lung accounts for only 2% of its body volume, compared with our 5%.

More importantly, however, birds have a unique system of thin-walled air sacs throughout the body, acting like bellows in conjunction with their lungs. The sacs are extensions of the bronchial walls (so are part of the respiratory system), but extend between the intestines, between the muscles, under the skin, even into some bones. There are usually nine of these very thin-walled sacs, though this can vary between species. Once the duck's air sacs are factored in, the effective lung volume rises from the modest 2% to an amazing 20%.

Largest are the two rear-most ones: the abdominal sacs. Air is drawn directly into them from the bronchi, by-passing the lungs. From there, it is pushed forwards through the lung into sacs at the front of the body, so air flows through a bird's lung only in one direction (very different from our diaphragm-driven 'in and out' system) and the whole contents of the lung are changed with each breath. This ensures that the blood flow in the lung is always going against the air direction, maximising oxygen uptake and carbon dioxide expulsion. Air reaching the front of the lung has already been depleted of oxygen as it moves forwards, but the blood it meets there has just arrived in the lung and it too is low in oxygen so is capable of taking more oxygen up from the relatively low air concentrations. As the blood moves back through the lung it becomes enriched with oxygen, but is meeting more and more highly oxygen-enriched air, so oxygen uptake is possible all the way through. This is a far more efficient system than our own.

Due to some remarkable detective work in 2005, we now know the origins of these sacs, a real key to birds' success (O'Connor and Claessens 2005; Ward 2006). Those origins lay in very low atmospheric oxygen levels around 230 million years ago (the reasons for that are probably a mix of biological and geological, and needn't concern us now, but such changes in atmospheric oxygen have been not unusual in Earth's history). They were about half of today's oxygen concentration at sea level, which equates to a modern altitude of around 4400 m. A lizard's lung is relatively small and rigid, which is fine when oxygen levels are high, but less so otherwise. Their ancestors doubtless struggled when levels dropped, and one minor group of reptiles came up with a radical solution. We know that hitherto unsuccessful group as the dinosaurs (and especially the theropods, from which the birds arose), and analysis of early dinosaur bones has shown that they developed the same type of cavities that modern birds have: in other words, they evolved air sacs to cope with the thin air. Their descendants – the birds – inherited them and developed them to assist with flight and survival at high altitudes, on mountains and above them.

Perhaps inevitably too, a few species have even adapted them for other uses. Male frigatebirds famously have a ludicrous great red throat pouch: an air sac just under the skin, which they inflate like a great balloon to attract females' attention. Moreover, the male is quite shameless in flaunting it to females flying over while he sits in apparent domestic bliss with his mate on the ground! (See Photo 18.)

Other memories of mountains

Like fleeting glimpses of snow-slashed black peaks, which loom through gaps in the swirling clouds, then withdraw into the diaphanous curtain, here are some random mountain memories:

- High in the Kapteinskloof Mountains north of Cape Town is a private nature reserve called Mountain Mist. I spent a memorable night there some years ago – the views alone are immense, north-east to the rugged Cederberg Mountains and west across the coastal plains 1000 m below to the Atlantic Ocean and Saint Helena Bay (reputedly the largest bay in Africa). As the sun sank over the sea, the distant strip-cultivated plains were charcoal and pale green and the sea was palest misty blue merging into the sky. The foreground tumbled away from my verandah in heath-covered rockiness ('fynbos' in the local usage). Proteas crowded in, and glowingly glorious Orange-breasted Sunbirds and taxonomically enigmatic Cape Sugarbirds worked the big flowers. The sugarbirds comprise just two species in a Family of utterly uncertain relationships: this one has a curved bill, brown back and white undersides streaked all over, and a hugely long tail. The sun swelled and turned that peculiarly African shade of vermilion just before disappearing. In the dusk, two sugarbirds sat up in silhouette. In the morning, the world on the mountain was in brilliant sunshine but below was a dense white blanket. I inched down the narrow rough road, sometimes through passages cut through the rocks, with mist stifling all views. Sometimes I saw a ghostly sugarbird by the road, once a silhouetted Ground Woodpecker, all in silence.

- Another misty day, near the summit of Mount Kosciuszko in Kosciuszko National Park, the highest point of Australia: this is an ancient land that has been eroding away for a very long time and 'Kosci' is just 2300 m above sea level. The wind ripped the cloud into shreds but it reformed immediately. It was mid-summer, but uncomfortably cold. A small flock of Little Ravens inspected the tumbled granite boulders, intently peering into crevices. They were looking for Bogong Moths: a medium-sized grey moth that emerges in spring on the black soil plains hundreds of kilometres to the north, where the larvae (known as 'cut worms') have spent the winter feeding on plant roots. As their 'foremoths' have done for millennia, the adult moths flew south at night, not feeding as they travelled, and spent the days somewhere dark and quiet, be it cave or broom cupboard. They find the same crevices that thousands of previous generations have used and spend summer in partial aestivation in the cool dark crevices. Little Ravens gorge on their fatty bodies, as do antechinuses (small marsupial carnivores). At night, inexplicably, some of the moths fly, though they still don't feed, and bats and Southern Boobook Owls are waiting. In autumn, the survivors fly north again, lay eggs – and die. No moth makes the trip twice: the map is hard-wired into their tiny brains. And each year the Little Ravens will be waiting for them.

- South of the 'gringo capital of the Atacama Desert', San Pedro de Atacama, high in the Chilean Andes at 4100 m above sea level is Laguna Miscanti: an almost impossibly beautiful lake, sky blue with the encirclement of snowy volcanoes reflected in it. The short walk to the lake is a challenge at that altitude, but eminently worth it. Andean Gulls, which live and breed only in the high Andes, drifted overhead. Scattered around the shore and on the water were massive Horned Coots, weighing over 2 kg, strangely attired with three black, tuft-tipped, fleshy wattles above the bill. In all the world they live only in these high Andean lakes more than 3000 m above sea level, in a small area around the corner where Chile, Bolivia and Argentina meet. The apparently floating nests we could see were an illusion: amazingly, the coots painstakingly carry pebbles out into the lake and drop them to make huge artificial islands on which they make their nest of vegetation. The world never ceases to astonish me (see Photo 19).

References

Averett N (2014) Birds can smell, and one scientist is leading the charge to prove it. *Audubon*, <http://www.audubon.org/magazine/january-february-2014/birds-can-smell-and-one-scientist>.

Balthazart J, Taziaut M (2009) The underestimated role of olfaction in avian reproduction? *Behavioural Brain Research* **200**(2), 248–259. doi:10.1016/j.bbr.2008.08.036

Birkhead TR (1974) Predation by birds on social wasps. *British Birds* **67**(6), 221–229.

Bonadonna F, Sanz-Aguilar A (2012) Kin recognition and inbreeding avoidance in wild birds: the first evidence for individual kin-related odour recognition. *Animal Behaviour* **84**(3), 509–513. doi:10.1016/j.anbehav.2012.06.014

Clark L, Avilova KV, Bean NJ (1993) Odor thresholds in passerines. *Comparative Biochemistry and Physiology. A. Comparative Physiology* **104**, 305–312. doi:10.1016/0300-9629(93)90322-U

Corfield JR, Price K, Iwaniuk AN, Gutiérrez-Ibáñez C, Birkhead T, Wylie DR (2015) Diversity in olfactory bulb size in birds reflects allometry, ecology, and phylogeny. *Frontiers in Neuroanatomy* **9**, 102. doi:10.3389/fnana.2015.00102

Downs CT, van Dyk RF (2002) Wax digestion by the Lesser Honeyguide *Indicator minor. Comparative Biochemistry and Physiology. Part A, Molecular & Integrative Physiology* **133**(1), 125–134. doi:10.1016/S1095-6433(02)00130-7

Elliott A (2016) Storks (*Ciconiidae*). In *Handbook of the Birds of the World Alive.* (Eds J del Hoyo, A Elliott, J Sargatal, DA Christie and E de Juana). Lynx Edicions, Barcelona, Spain, <http://www.hbw.com/node/52205>.

Elphick J (2014) *The World of Birds.* CSIRO Publishing, Melbourne.

Forshaw J (2002) *Australian Parrots.* 3rd edn. Alexander Editions, Robina, Queensland.

Houston DC, Copsey JA (1994) Bone digestion and intestinal morphology of the Bearded Vulture. *The Journal of Raptor Research* **28**(2), 73–78.

Ksepka DT (2014) Flight performance of the largest volant bird. *Proceedings of the National Academy of Sciences of the United States of America* **111**(29), 10624–10629. doi:10.1073/pnas.1320297111

Langston NE, Rohwer S (1996) Molt–breeding tradeoffs in albatrosses: life history implications for big birds. *Oikos* **76**, 498–510. doi:10.2307/3546343

Lee SI, Kim J, Park H, Jabłoński PG, Choi H (2015) The function of the alula in avian flight. *Scientific Reports* **5**, 9914. doi:10.1038/srep09914

Margalida A (2008) Bearded Vultures (*Gypaetus barbatus*) prefer fatty bones. *Behavioral Ecology and Sociobiology* **63**(2), 187–193. doi:10.1007/s00265-008-0649-6

Negro JJ, Sarasola JH, Farinas F, Zorilla I (2006) Function and occurrence of facial flushing in birds. *Comparative Biochemistry and Physiology. Part A, Molecular & Integrative Physiology* **143**(1), 78–84. doi:10.1016/j.cbpa.2005.10.028

Nevitt GA (2008) Sensory ecology on the high seas: the odor world of the procellariiform seabirds. *The Journal of Experimental Biology* **211**, 1706–1713. doi:10.1242/jeb.015412

Nevitt GA, Bonadonna F (2005) Seeing the world through the nose of a bird: new developments in the sensory ecology of procellariiform seabirds. *Marine Ecology Progress Series* **287**, 263–307.

O'Connor PM, Claessens LPAM (2005) Basic avian pulmonary design and flow-through ventilation in non-avian theropod dinosaurs. *Nature* **436**, 253–256. doi:10.1038/nature03716

Rajchard J (2008) Exogenous chemical substances in bird perception: a review. *Veterinarni Medicina* **53**(8), 412–419.

Reddy G, Celanib A, Sejnowskic TJ, Vergassolaa M (2016) Learning to soar in turbulent environments. *Proceedings of the National Academy of Sciences of the United States of America* **113**(33), E4877–E4884. doi:10.1073/pnas.1606075113

Rohwer S, Ricklefs RE, Rohwer VG, Copple MM (2009) Allometry of the duration of flight feather molt in birds. *PLoS Biology* **7**(6), e1000132. doi:10.1371/journal.pbio.1000132

Sanz JL, Chiappe LM, Pérez-Moreno B, Buscalioni ÁD, Moratalla JJ, Ortega F, *et al.* (1996) An Early Cretaceous bird from Spain and its implications for the evolution of avian flight. *Nature* **382**(6590), 442–445. doi:10.1038/382442a0

Savoca MS, Wohlfeil ME, Ebeler SE, Nevitt GA (2016) Marine plastic debris emits a keystone infochemical for olfactory foraging seabirds. *Science Advances* **2**(11), e1600395. doi:10.1126/sciadv.1600395

Sievwright H, Higuchi H (2016) The feather structure of Oriental Honey Buzzards (*Pernis ptilorhynchus*) and other hawk species in relation to their foraging behavior. *Zoological Science* **33**(3), 295–302. doi:10.2108/zs150175

SouthAfrica.net (2016) *Lord of the Rings* writer has South African roots. SouthAfrica.net, Johannesburg, South Africa, <http://www.southafrica.net/za/en/articles/entry/article-southafrica.net-jrr-tolkien>.

Sterbing-D'Angelo S, Chadha M, Chiu C, Falk B, Xian W, Barcelo J, *et al.* (2011) Bat wing sensors support flight control. *Proceedings of the National Academy of Sciences of the United States of America* **108**, 11291–11296. doi:10.1073/pnas.1018740108

Tucker VA (1968) Respiratory physiology of House Sparrows in relation to high-altitude flight. *The Journal of Experimental Biology* **48**, 55–66.

Ward PD (2006) *Out of Thin Air: Dinosaurs, Birds, and Earth's Ancient Atmosphere.* National Academies Press, Washington DC, USA.

Ward J, McCafferty DJ, Houston DC, Ruxton GD (2008) Why do vultures have bald heads? The role of postural adjustment and bare skin areas in thermoregulation. *Journal of Thermal Biology* **33**, 168–173. doi:10.1016/j.jtherbio.2008.01.002

Photo 1. Red soil and termite mounds and polished gibbers: desert country around Warrigal Waterhole, north-west Queensland. (Page 1)

Photo 2. Courting Budgerigar pair, Mullewa, Western Australia. A budgie can breed at just 6 weeks old. (Page 8)

Photo 3. Ethiopian Swallows, Waza National Park, northern Cameroon, panting to stay cool in 40 plus degree heat by evaporation from the open bill. (Page 17)

Photo 4. American Flamingo, Isabela, Galápagos, showing typical flamingo feeding strategy: pumping water in and out of the upside-down bill, filtering out tiny food items. The rich plumage colour is obtained solely from this food. (Page 23)

Photo 5. Primary lowland tropical rainforest in Korup National Park, south-western Cameroon: the mighty granitic massif of Rengo Rock gives a vantage point. (Page 27)

Photo 6. Golden-naped Barbet, Mount Kinabalu National Park, Sabah. The yellow and the blue are achieved by completely different methods, and the green by a combination of them. (Page 36)

Photo 7. Ocellated Tapaculo, Paz de las Aves, north-western Ecuador: a near-mythically hard-to-see dweller of the Andean cloud forests, patiently habituated here to come to offered worms. (Page 39)

Photo 8. Wire-crested Thorntail hovering: one of the most dramatic of hummingbirds (a big call indeed!). This one was attending the feeders at Aguas Verdes, eastern Andean slopes, northern Peru, which is another ecotourism success story. (Page 41)

Photo 9. Rufous-bellied Euphonia calling 45 m above the ground, seen from a canopy tower in a huge Kapok Tree, Sacha Lodge, Ecuadorian Amazonia. (Page 53)

Photo 10. Medium and Large Ground Finches scrounging, Baltra Airport, Galápagos: their different-sized bills evolved to allow them to co-exist by eating different-sized seeds – though not from plates! (Page 55)

Photo 11. Common Cactus Finch, Puerto Ayora, Galápagos. The cactus finches have evolved as complete cactus specialists, eating flowers, fruit, stems and the insects on them, using their long sharp bills. (Page 60)

Photo 12. Tasmanian Nativehen, Copping, Tasmania: the 'turbo chooks' have lost the use of their wings and rely on their long strong legs for safety. (Page 64)

Photo 13. The Titicaca Grebe is flightless and restricted to huge Lake Titicaca in the high Andes of southern Peru. In this case, the lake is effectively an island in the mountains; retaining flight would have made the bird susceptible to being blown out to die in the hostile montane landscape. (Page 69)

Photo 14. Flightless Cormorant pair at seaweed nest, Fernandina, Galápagos. The wings are clearly far too small to allow flight. One bird has just returned from fishing and they are performing a greeting ceremony. (Page 70)

Photo 15. Red-tailed Tropicbird, soaring in courtship display off the wild Malabar Cliffs on the northern edge of Lord Howe Island. (Page 75)

Photo 16. Andean Condors soaring up out of Colca Canyon, southern Peru, where they roosted for the night; they are riding rising air currents as the morning warms. (Page 89)

Photo 17. Red-capped Parrot, Albany, Western Australia, where it is endemic. The elongated upper mandible, used for extracting tiny seeds from big Marri eucalypt capsules, is evident. (Page 97)

Photo 18. Magnificent Frigatebird pair at nest, North Seymour, Galápagos. The male has inflated his preposterous red throat sac to impress both his mate and other females flying over; it evolved from one of the air sacs that make bird respiration so efficient. (Page 101)

Photo 19. Laguna Miscanti, at 4100 m above sea level in the Atacama Desert of northern Chile, supports populations of birds including the huge, and very restricted, Horned Grebe. (Page 103)

Photo 20. A pair of Waved Albatrosses in a breeding colony on Española, Galápagos, perform an elaborate bill-clashing bond reinforcement ceremony as one returns from a fishing expedition. Their sole chick is the object of intensive care. (Page 133)

Photo 21. Cocha Salvador, deep in the primary lowland rainforest of the Manu Reserved Zone in southern Peruvian Amazonia, is an oxbow lake, formed from a cut-off bend in the Manu River. In a morning there, we saw four birds that are the only members of their Families. (Page 134)

Photo 22. The Limpkin, here at Cocha Salvador, is one such bird, a single-species Family. It is not at all related to the ibises, which it superficially resembles. Its calls have been used in movies from Tarzan to Harry Potter! (Page 137)

Photo 23. Hoatzin, Tambopata National Reserve, southern Peruvian Amazonia. This remarkable bird is the only member of an entire Order, one of only two such birds in the world. It is also the only bird that has evolved colonies of gut bacteria to digest cellulose. (Page 138)

Photo 24. Yellow-billed Spoonbill, Jerrabomberra Wetlands, Canberra. This bird is preening: running each feather in turn through its bill to clean it and 'rezip' the barbules. This is an immense task, which every bird undertakes every day. (Page 146)

Photo 25. Pied Kingfisher, Queen Elizabeth National Park, Uganda. This is one of many bird species in some 30 Families that breed in excavated burrows. (Page 173)

Photo 26. Collared Sparrowhawk with House Sparrow, suburban Duffy, Canberra. This little drama took place in our small backyard: it's amazing what happens in your garden! (Page 177)

Photo 27. Gimlet (*Eucalyptus salubris*) woodland, near Norseman, inland southern Western Australia: such woodlands are biologically very rich. (Page 181)

Photo 28. Darwin's Rheas, Torres del Paine National Park, Chilean Patagonia. These are ratites: flightless, mostly large, southern birds (the ancestral rhea flew into South America long ago). The male is the sole carer, and they produce many young to increase the chance that some will survive. (Page 183)

Photo 29. Mealy Parrots, Blanquillo clay lick, southern Peru. The 'mealy' (i.e. floury) look is due to an abundance of powder down, a talc-like powder produced by highly specialised decomposing feathers to assist in general feather care. (Page 189)

Photo 30. Nankeen Kestrel, wind hovering near Canberra: balancing itself against the wind by constantly adjusting wings and tail. The alulas (special feathers on the 'elbow' of the wing of every flying bird to break surface turbulence and greatly improve aerodynamics) are quite visible here. (Page 206)

5
Wetlands and rivers

Gum Swamp, New South Wales

There was something chillingly surreal about that hunt. The hunter sank in the water until not much more than eyes and nostrils were visible, then edged forward slowly with barely a ripple. The effect was eerily crocodile-like. The mother of the intended prey suddenly became anxious, then stood up in the water with flapping wings, yelping loudly and harshly, at which the eight Grey Teal chicks rushed across the surface, either towards her or the nearby reeds. At that moment the hunter also made his own rush across the water surface; when the roiling water and terrified chaos had subsided, the eight chicks were now seven. The male Musk Duck was not to be seen until he surfaced some 30 m away, a limp bundle of downy feathers in his beak.

This little drama played out at Gum Swamp (and isn't that an archetypal Australian place name?) on the edge of the historic gold-mining and pastoral town of Forbes in the south-western slopes of New South Wales. In general terms, this swamp is in many ways fairly typical of wetlands in modern southern Australia, and indeed beyond. It comprises several hectares of shallow open water fringed with reeds and rushes, with an adjacent area of somewhat degraded woodland. It has a rudimentary, but functional, concrete bird hide, which is by no means a universal accoutrement at such sites. However, the giveaway to the recent nature of its current state is the large number of standing dead eucalypts in the water. This may once have been an ephemeral wetland – in which case the drowned trees are probably River Red Gums *Eucalyptus camaldulensis* – or it could just have been woodland, but it is now artificially permanent, fed constantly by an outflow of treated water from the adjacent sewage works. There are very few significant wetlands outside the Australian tropics in anything like their original condition. In this land of El Niño, this means ephemeral: filling in the rains, especially of the wet La Niña seasons, and gradually drying as the months and years go by with evaporation exceeding rainfall and subsequent inflow. (Mind you, given our record in draining and choking off natural wetlands, I'm happy for any compensation going in the form of artificial ones, but we also need 'real' ones.) The dead trees were mature when they drowned, because many have nesting hollows. Cockatoos use some, but they also support a huge population of Feral Pigeons, which in turn attract regular visits from Peregrine Falcons, a few from Black Falcons, and occasional ones from the mythically rare Grey Falcon (never while I've been there though!). Scores of ducks of up to nine species are scattered across the waters, along with grebes, moorhens and coots. Herons and rails patrol the edges where Australian Reed-Warblers and Little Grassbirds sing from the reeds, and ibis and spoonbill perch in the dead trees.

r-selected breeders: 'have lots of kids, hope a couple survive'

The little Grey Teal family tragedy is a daily occurrence in the lives of duckling broods, which is precisely the reason that there are so many eggs in the clutch, and so many youngsters hatching from them (minus the odd egg that falls to an opportunistic goanna or snake or raven).

In a dead tree across the swamp from the hide is a huge stick nest belonging to a White-bellied Sea Eagle, which has been there for generations and may weigh hundreds of kilograms. They generally have only one chick (if two hatch, the second rarely survives to leave the nest) on which they lavish all their attention, investing hugely in its survival. This, as we've discussed elsewhere (page 87), is referred to by ecologists as a K-selected breeding strategy: it is an adaptation to steady conditions over a long period of time. Indeed, K is a mathematical symbol referring to the carrying capacity of the local environment. It has been taken for this usage from a complex equation (complex by my, and possibly your, standards at least!), which was derived in 1838 by Belgian mathematician Pierre François Verhulst to model population changes. K-selected breeding species, such as the Gum Swamp sea-eagle, rely on a stable world in which they maintain their generally low population at a level close to what the environment can support.

The teals, however, go for 'the other' option: they are r-selected breeders. In Verhulst's equation, r stands for the maximum rate of increase of the population. (I'm sure someone could tell you why Verhulst used a capital K and a lower case r, but I'm afraid that person is not me.) Such organisms have many offspring, with minimal attention to each one, and rely on chance and basic care (or in some cases none at all) to ensure that one or two get through. In an unpredictable environment – such as that provided by a wetland in an El Niño-driven climate – r-selected species with their high rates of population growth and generally unfilled niches are at an advantage. Strictly speaking, if a pair of birds throughout their life only manages to have two offspring that grow up to reproduce they would be breaking even, though it's likely that they would be behind some of their neighbours in the genetic competition. The eight ducklings the mother teal was trying to protect represents the mean clutch size for the species. Others such as the Australian Wood Duck can have up to a dozen eggs, and the peripatetic Plumed Whistling Duck, which occasionally turns up at Gum Swamp from northern Australia, can have up to 14. Actually, larger clutch sizes are regularly reported, but many ducks have a tendency, especially in drier seasons, to dump eggs in someone else's nest to be looked after. The nest of a Red-crested Pochard was reported to contain 39 eggs (Carboneras and Kirwan 2017), most of which she had certainly not laid!

Parasitic ducks

Digressing for a moment – though can a good story ever really be digressive? – the Black-headed Duck of central South America is an obligate nest parasite: the only duck that never builds its own nest. Amazingly, its hosts include not just other ducks, but coots and gulls are important too; other hosts include ibis and even raptors such as Snail Kites and Chimango Caracaras (the latter being an egg predator!). These ducks are not like cuckoos, demanding to be fed: after hatching, the duckling slips out of the nest and disappears, already self-sufficient. It means that the mother can lay as many eggs as she can produce, not just as many as she can brood.

However, the real r-selected specialists among vertebrates are fish, which can lay thousands of eggs. Beyond that, insects can have huge broods and plants such as eucalypts can produce hundreds of thousands of seeds.

Nature, as ever, refuses to sit neatly in the boxes we create for her: they are for our benefit not hers. For this reason, it's important to remember that it isn't just a question of a bird being either r-selected or K-selected: those are just an indication of the extremes. We could almost certainly find (if we had that much spare time to comb determinedly through the literature) species with mean clutch sizes of every number between the sea-eagle's one and the teal's eight. Moreover, the numbers themselves don't tell a uniform story: for instance, the White-bellied Sea-Eagle produces one egg (occasionally two) a year, but the Andean Condor lays just one every 2 years.

For the most part, our little brains tend to prefer simple definitions without too much nuance. Bad luck then when our brain is forced to consider a situation such as that of the Letter-winged Kites of the arid desert plains and grasslands of the Lake Eyre Basin and Barkly Tablelands of inland Australia. In 'normal' (i.e. dry) seasons, they behave like we expect a raptor to: producing one or two young a year, or none at all in drought times. Then eventually La Niña comes and the rains reach deep into the centre of the continent, flooding vast areas and isolating homesteads and communities for weeks at a time. As the waters recede, the land comes pulsingly alive with unimaginable bounties of grass seeds, flowers and insects and the birds and other vertebrates that feed on them. At this time, the Long-haired Rat *Rattus villosissimus* breeds in vast numbers – and the Letter-wings' time has come. With their main food supply now assured, they cast caution to the desert wings, and flip from being a 'normal' K-selected raptor to a classic El Niño r-selected bird, producing up to six young at a time. When the land dries again, and the surviving rats retreat to the fissures in the now-cracking black soil plains, the young kites disperse across the country, many dying hundreds or thousands of kilometres from the land of their 'chickhood'.

'Normal' really is such a dangerously alluring and comfortable concept to rely upon.

Grey Teal and breeding triggers

Indeed, when a remarkable man called Harry Frith turned his attention to Grey Teal (and other waterfowl) in the 1950s, he refused to accept that these children of El Niño were 'normal' in the context of Northern Hemisphere understandings of how breeding triggers work.

Frith had a degree in agricultural science, but his career was interrupted before it had begun, like that of so many others, by the Second World War. He fought in the Middle East and New Guinea, then returned to work in the citrus orchards of the Riverina in southern New South Wales. His heart wasn't in it, however, and he switched to the Wildlife Survey Section of the Commonwealth Scientific and Industrial Research Organisation (CSIRO), the peak government research body. He taught himself ecology and zoology to such good effect that a decade later he was appointed the first chief of the new Division of Wildlife Research. Frith was instrumental in our understanding of Australian waterfowl and the Malleefowl, and was a great conservationist who played a key role in the declaration of Kakadu National Park in the tropical Northern Territory: one of the world's great parks (Tyndale-Biscoe *et al.* 1995). (We may see it as ironic that very many ducks were killed in the cause of conservation, but it was essential to understand what was happening, especially with regard to breeding status and preferred food supplies. The numbers of these deaths were as nothing compared with the impacts of habitat loss and competition from agriculture and other sectors of the economy.)

In strongly seasonal climes, changing day length triggers the production of hormones that start the courting and breeding cycle at the same time every year, but Frith saw that that wouldn't work here – it might start waterbirds breeding when there was no water and prevent them from doing so when conditions were perfect. It wasn't rainfall *per se*, as he observed that that alone didn't trigger a reaction. What he observed was that the *sight* of rising water sparked Grey Teal breeding behaviour in anticipation of the food bonanza that the birds 'know' will come. They began displaying within a day or so of this happening, and ovulation began soon afterwards, well before a change in food could be the stimulus. If water rose, then fell again, displays petered out and there was no ovulation (Frith 1967; Andrewartha and Birch 1986).

I recall my excitement as an impressionable young undergraduate when I heard that story from another giant of Australian ecology, HG Andrewartha at Adelaide University.

The members of that little family of Grey Teal whose drama I witnessed could have ended up anywhere in Australia if they survived. A waterbird in inland Australia must move when the waters dry up, or perish. Nationally coordinated banding studies – overseen by Frith – showed that, when conditions are tough in the Murray–Darling Basin, as happened in 1958, Grey Teal scattered throughout the entire continent, and as far as New Guinea and even New Zealand. They fly fast and hard in straight lines; if they fail to find water they either reach the coast or perish, as very many do. This is a variation on the nomadism that is key to the survival of inland Australian birds (see Chapter 1).

Musk Ducks: a story of a duck's luck

The more we learn about the southern Australian Musk Duck, the more we realise both that it's worth knowing and how much there is to know. At Gum Swamp I've seen not only the hunt – a very rare behaviour among duck species, most of which are vegetarian as adults – but have also seen a female Musk Duck carrying two chicks on her back. This is also uncommon among ducks, though not for swans. We used to be fairly comfortable in assigning the Musk Duck to a place with the stifftail ducks: a group comprising a genus of six species, including the Australian Blue-billed Duck, which between them are found on every unfrozen continent, plus a couple of South American species in separate genera. Now, however, we tend to cough a bit uncomfortably and suggest that we never really thought that the Musk Duck belonged there ...

The problem is, we don't really know where it does belong. Maybe its ancestor was one of the stifftails back in a long-ago Gondwanan lake, and isolation has led it down different paths. Or, perhaps more likely, it really is an old Gondwanan with no meaningful foreign relatives, other than perhaps the equally enigmatic Australian Pink-eared Duck (there is some genetic evidence for that) and resembling the stifftails simply because of its similar freshwater diving lifestyle.

It is one of the few birds named for its smell. Not many individual birds can claim (albeit posthumously) to have unequivocally given their name to the entire species. That, however, is the dubious consolation for an unfortunate male Musk Duck, cut down not only in his prime but at the most inconvenient time of year for him. We know this, not only because he was collected (in 1791 by Archibald Menzies, inland from current-day Albany) at the height of his spring breeding season, but because his post-mortem revenge was to imbue the entire ship with his courtship perfume. This, it seems, had been more alluring to female Musk Ducks than it was to male British sailors. The drake exudes the chemical with his preening oils from his uropygial gland when he most needs to impress his intended. As far as I know, not many birds employ this strategy, which is more often associated with mammals or insects.

However, it wasn't just the late duck whose luck ran out; Menzies too was badly short of this commodity. George Vancouver (a veteran of Cook's Pacific expeditions) was sent by the Admiralty in the *Discovery* in 1791 to sort out the Spanish who were being a nuisance off the north-west coast of North America. As ever, unofficial British biologist-in-chief Sir Joseph Banks made sure there was a scientific presence, in the person of Archibald Menzies, a naval surgeon and botanist. At that stage, the west coast of North America was about as far from England as you could get. If you didn't want to go via Cape Horn (and who in their right mind would?!) the option was the Cape of Good Hope, east across the Southern Ocean and north-east across the entire Pacific. Vancouver decided that he might as well have a look around while he was in the neighbourhood of south-western Australia. Menzies used his time industriously: he gathered 'wild celery' to counteract scurvy, collected many native plants and planted vine cuttings and watercress, and almond, lemon, orange and pumpkin seeds. (The plantings, unsurprisingly, did not survive.) And, in addition to the Musk Duck, he collected specimens of (and wrote in his journal the first descriptions of) the Western Rosella, Southern Boobook and Red-capped Parrot. Sadly he didn't ever publish, so others later got the credit. It got worse for him though – he fell out with Vancouver and got locked up on board for 3 months, during which time his carefully collected and tended live plant collection perished. The duck was probably smiling grimly to himself at the time.

Showing off in the water

Until his sudden and unfortunate demise, the duck would have participated in one of the most unusual and impressive courtships of any waterbird. Swimming out from the fringing reeds, he kicked powerfully with one foot at a time, each stroke hurling a spout of water behind him. When he got to the open water he stopped, while still hurling the water out behind him (which in itself seems to be a pretty clever trick). With each kick, he raised his wings to meet above his back. This continued for some minutes, while his whole appearance changed. The fleshy lobe beneath his bill distended with blood while he puffed out his neck and checks so that his raised head resembled a ball. At the same time, he lifted his tail and fanned the stiff tail feathers across his back. By now he barely resembled a duck. But just in case no-one had noticed yet (very unlikely), he let the world know in no modest terms, with a series of deep resonant 'ker-PLONKs' echoing across the water from his feet. (Humans haven't really worked out how a Musk Duck does this.) Finally, the whole performance reached a crescendo as he sank in the water, slowly rotating while still hurling gouts of water out behind him, his bill pointing to the sky as he whistled loudly and shrilly for minutes at a time. Such a display tends to attract an audience of both males and females, understandably enough. A successful and brief mating with one of the female admirers marks the end of the spectacle.

This is by no means the only water-based display, however, and why would a waterbird not choose its major habitat as its stage?

Cocha Salvador, Peruvian Amazonia

Cocha Salvador at dawn is a special place to be. The mist rises from the surface of the big deep lagoon and primary rainforest crowds down to the shores. Across the water rolls the essential sound track of Amazonia: the pulsing wind-like roar of the howler monkeys.

It's not an easy place to get to. We had travelled for 2 days in a motorised narrow river boat from the frontier river port of Atalaya

Other waterbird dancers

In the Sacred Valley of the Incas in the Andes near Cusco in southern Peru, the Urubamba River rushes cold and ice-green beneath mountains scored by ancient agricultural terraces, built and worked before the Incas imposed their rule on the valley. The water foams and crashes furiously onto the huge boulders in the stream, hurling itself in boiling torrents through channels between the rocks. Clearly nothing of any size could survive in such a maelstrom. And yet …

One morning I walked from our riverside hotel to the edge of the little gorge through which the Urubamba cascades – and saw something remarkable. Three slender long-tailed ducks were standing on the rocks, water racing over their feet. Two were dark-bodied with white heads slashed with a black line from the red bill through the eye to the base of neck; the third was dark above and brick red below. The white-heads were whistling loudly, beaks held high, the shrill calls cutting through the river's roar. I knew they were Torrent Ducks: specialists in the wild Andean streams along the full length of South America, from close to the Caribbean down to Tierra del Fuego. I was surprised though: Torrent Ducks are noted as faithful partners for as long as they both live and real home-bodies, fiercely defending a territory which stretches for a kilometre or so along the river. Nonetheless, here were two males (the white-heads) clearly competing for the female's favours. She was watching the whistling duel with some interest, but was in no hurry to make a decision.

Then, startlingly, they were all in the maelstrom, swimming strongly upstream! And from I knew not where, a third male suddenly appeared and joined the competition. The three males led the way, whistling constantly and showing off their power and skill in the horrific conditions, while she followed them, still presumably taking notes. This enthralling performance continued, on rocks and in the water, for some time until we finally had to leave. I can only assume that she had recently lost her mate and that the others were singles cruising and looking for such an opportunity; it is reported that males at least usually find a new mate within a week of losing one (Cadona and Kattan 2010).

Perhaps the most famous of the water dancers, however, are the grebes: a truly venerable and worldwide Family of divers with apparently no close relatives. An important difference between their displays and those of the ducks I've described is that only the male ducks perform while the females watch and judge. But in grebes both members of an already mated pair participate equally, often swapping roles. Many years ago I was working at the Coorong, a hugely atmospheric long narrow lagoon stretching south from the mouth of the Murray River in South Australia for 150 km and separated from the wild Southern Ocean only by a strip of sand dunes. I found a gap in the reed beds and crouched and peered through, just in case … Good move!

I was mesmerised by what I saw. A pair of magnificent Great Crested Grebes was only 30 or so metres away on the water, entirely engrossed with each other.

These are big birds: up to 60 cm long and weighing up to 1.5 kg, with a long neck and straight sharp fish-snatching bill. The body, generally low in the water, is dark and unremarkable – but the head! The black forehead rises to a spiky black erectile crest, while below that a white face surrounds a red eye. The rich rufous sides of the head sweep back to a glorious mane of black-tipped chestnut feathers. In the Northern Hemisphere they lose these adornments in winter, but in Australia and Africa (this is a very widespread species – or perhaps a species group – absent only from the Americas) they remain magnificent all year round.

The two Coorong birds were sitting in the water facing each other, in silence, turning their heads from side to side with raised crests, sometimes bending their necks backwards so that their heads touched their backs to rearrange some feathers, from time to time quickly shaking their heads or nodding them to fan their manes out to show their full glory. After a few minutes it seemed as though it was all over as they turned and swam away from each other, then slowly sank beneath the choppy surface. Suddenly one, then the other, resurfaced with a beakful of water weed. They swam right up to each other, then rose in the water breast-to-breast, rocking their heads from side to side in unison, paddling furiously to stay erect. I was breathless, so I can't imagine how they were feeling! I've since learnt that this is called the 'weed ceremony' and is practised by Great Cresteds all over the world, but it was a revelation to me at the time (and I've never seen it since).

The closely related Western and Clark's Grebes from North America have an element to their display which is, if anything, even more dramatic. After some head-shaking foreplay, a message is exchanged by a quick look between them, then they rise completely out of the water so that body and neck are vertical, with head held stiffly forward and wings held stiffly out behind. Astonishingly, they rush across the water in a way that seems to suspend normal laws of physics – they run on the surface. They do it via the expenditure of huge amounts of energy, pattering their splayed webbed feet at something like 20 little steps a second on the surface, keeping perfect pace side by side with each other. After a few seconds, and some 20 m, they perform a synchronised dive and disappear.

But why? In the case of the ducks, it's relatively obvious. The male is showing off his fitness and, I suspect, his capacity as a survivor – it must take years to perfect the Musk Duck's complex rituals and, as well as demonstrating his strength and resilience, he is declaring that he has genes that have enabled him to live a long life, no small advantage. Those are the genetic edges that any discerning female Musk or Torrent Duck wants for her offspring.

But what about the grebes' elaborate dance, which is apparently mostly performed by already established couples? Here, it seems to me, we don't really have firm answers, though there are some assertions regularly repeated. I am bearing in mind that displays such as those of the grebe pairs are immensely expensive in energy, in time (which could be perhaps better spent in feeding or nest-building, for instance), in production of otherwise unnecessary display plumage and in the risk of attracting predators. It is often suggested that such behaviour enables birds of similar species to ensure they're not making a very embarrassing mistake in their choice of mate. There are such similar species pairs, including Clark's and Western Grebes, but surely any confusion can be only in our eyes? I would be very surprised (though in fact I often am!) if Clark's and Western Grebes really needed to undertake complex recognition tests to know if a bird was of their own species or not. And wouldn't a simple vocalisation be a lot more efficient if such a test was required? In any case, that is certainly irrelevant to the Great Cresteds, which resemble nothing else.

Moreover, it's a red herring if the pair has already mated. It is said that displays in monogamous species (which may mean a lifetime pairing, or just for one season) are to re-establish links, though it's not clear why that is necessary if breeding has already commenced. Is it necessary to periodically check on your mate's continued commitment? Or their continued fitness? Well, perhaps. Life-bonded birds such as albatrosses undertake elaborate displays, especially at the nest, with song and bill-clattering dance (see Photo 20). In their case, they are taking it in turn to forage at sea for extended periods while the other broods the egg. Could it be that the display assists in assessing the fitness of the stay-at-home partner to stay on the nest without feeding, so that the forager can judge how long it's safe to be away? That sounds a bit far-fetched, but I think it's plausible.

That can't explain the grebes' dances though. Moreover, it seems that most grebes form a pair bond only for one season. Is it possible that in their case at least we have consistently misunderstood and that the dances are mutual tests of suitability to share this season's reproductive adventure, rather than reaffirmations? Probably not, but it would at least be easier to understand if that were the case. Perhaps it is a process – an extended courtship during which bonds are gradually formed – which only ends when eggs are laid. If so, that might explain why it is unclear from the written accounts if the dancing pair are 'an established item', or only just becoming one, but that really only begs the bigger question. Why would such an expensive and exhausting process be necessary, or even advantageous?

In all the accounts I've been able to find, these questions seem to be skipped over – though of course it could just be that I'm missing something obvious; it would hardly be the first time! However, far from being depressed by my ignorance, I revel in all the things that I don't know and, as in this case, even those that we collectively don't seem to know. That is not to say that I think that ignorance is bliss – I will lie awake gnawing at a question that I can't think of a plausible answer to – but I do love being reminded of how much we in general (and me in particular) have to learn. It would be a grim day indeed, for a couple of reasons, if I were to awake one morning and decide there was nothing left for me to discover! *Inter alia*, my ignorance helps me remember just how trivial I am in the overall scheme of things, and that's a most important reminder.

down the Alto Madre de Dios River, then up the Manu River to the basic lodge run by the Machiguenga people. This morning, in the dark, we got back into the boat to travel downstream again, then walked through the gradually lightening forest to a little jetty, where we clambered onto a heavy rough-hewn raft that our indomitable bird guide Ribelino poled slowly across the water.

Cocha Salvador is a very large oxbow lake (a billabong in Australia): a former great bend of the Manu River cut off by floods and now forming a deep still backwater draped with rainforest (see Photo 21). It is in the Manu Reserved Zone: a vast lowland rainforest wilderness within the Manu Biosphere Reserve, inhabited by indigenous people and only otherwise accessible to researchers and visitors accompanied by authorised and environmentally trained guides. Cocha Salvador is in this reserve, not in Manu National Park as often claimed in websites of companies who go there – the park itself is even further upstream, and is closed to all visitors except authorised researchers. We had the lake to ourselves; only one group is allowed on it at a time, and Ribelino had ensured that we got the prized first turn on the water.

One of our main 'targets' for the morning was one of the most impressive, and rarest, big mammals in South America. The big oxbow lakes – and they are few – are key habitats for Giant Otters, an endangered species across their northern Amazon Basin range. Heavy hunting for skins has reduced their numbers to no more than 5000; the species is listed as Endangered by the International Union for Conservation of Nature. Even in remote Manu, it is estimated that only a dozen families survive, and one of these is in Cocha Salvador. We were enthralled by our encounters with the family of huge otters – up to 1.8 m long and weighing up to 30 kg (though much larger animals were reported before the days of intense hunting). They share the cocha (from a Quechua word, the language of the Inca empire) with their mortal enemies, the Black Caiman, largest of the alligators, which can grow to 5 m long. Both otters and caiman prey on each other's youngsters; the otters will also team up to attack larger caiman. We watched the otters bring big fish ashore to share, albeit tetchily, all the while squealing and whistling and whining.

It was thrilling and engrossing, but this is a book about birds and something else remarkable was happening at the cocha; no fewer than four single-species bird Families were present!

Single-species Families

Any birder is interested in a bird species which is the only member of an entire Family (birds with no close relatives, and for which it is often unclear just what even their more distant relationships are). They don't come along every day; there are something like 10 500 living species of birds, and only 30–35 (depending on which taxonomy you use) of them are alone in their Family. Tools, especially biochemical ones, are improving all the time and new ones are being developed, but ultimately it still requires the experience of a scientist (or more usually a team) to make a judgement as to whether the differences revealed in the chromosomes or the proteins of two birds represent evidence of one or two species, or even Families.

New tests can reveal unsuspected differences. Until very recently, the apparently unremarkable little Spotted Wren-babbler of the dense mountain thickets of eastern India, China and South-East Asia wasn't even regarded as in its own genus, but was included with eight other species of Asian wren-babblers in the large Old World babbler Family. Then a major in-depth study of a very large number of mostly Old World passerines, published in 2014, threw up the totally unexpected result that the unassuming Spotted Wren-babbler was royalty, having its origins way back in the early days of passerine evolution and losing any relatives long ago; it has an entire Family all to itself (Alström *et al.* 2014). To reflect this, it is now known as the Spotted Elachura, which is also its new genus name.

The reverse can also happen, for a couple of different reasons. First, in a mirror image situation of the Elachura's, a bird previously thought to be unique can be reinterpreted as more closely related to another bird group than was believed. In the case of the Sapayoa of Panama and northern South America, many eminent people now think it to be more closely related to another bird group than was believed even possible (e.g. Moyle *et al.* 2006)! The highly influential International

Ornithological Congress (IOC) believes that the Sapayoa belongs with the Old World broadbills: a Family of primitive suboscine passerines (see page 46). This would make it the only American member of the Old World suboscine grouping, which includes broadbills, pittas and the Madagascan asities. (Not for nothing is its species name *aenigma*!) Not everyone is convinced by this, and some would retain its single-species Family status, but it illustrates the point.

The other way in which a single-species Family bird can lose its coveted status (coveted by birders, if not actually the bird) is if it is revealed that two species have actually been masquerading as one. This is the case with the Ostrich, for instance: most ornithologists now think that the Somali Ostrich of the Horn of Africa is a separate species from the more widespread Common Ostrich. Likewise, the IOC would split the Osprey, raising the Eastern Osprey of Indonesia and Australia to full species status, thus robbing its Western relatives of their special standing.

However, the four Robinson Crusoe species (i.e. all on their own) that we saw that morning at Cocha Salvador seem to be in no danger of losing their status, though some would include the Donacobius with the wrens. Even in this case, though, the main proponent of the stance among the major world taxonomy players, the highly respected *Handbook of Birds of the World*, acknowledges that the evidence is shifting away from it. Let's start there.

The Donacobius (or Black-capped Donacobius, though there are no others from which it needs to be distinguished) is a medium-sized passerine with a long tail, not unlike a mockingbird (with which it has also been included). Dark above and rusty below, they sit in pairs or small groups high in fringing rushes and reeds and whistle at passers-by on the water. They are found around waterways throughout much of lowland northern South America.

The other three were all non-passerines. Of these, the Limpkin looks perhaps the most 'normal': a bit like a brown ibis with a black and white speckled neck and a slightly curved bill. Its nearest relationship seems to be with the cranes. It prowls the water's edge with a distinctive high-stepping gait (which allegedly gives rise to the name, though it doesn't really resemble limping), searching for the big snails that are the mainstay of its diet. It deals with them in a very similar way to that of the openbill storks (see page 98): an excellent example of parallel evolution. Its wonderful wild harsh squealing calls appeared in many Tarzan movies (set, of course, in Africa), and most recently they gave their voice to the mighty and chimeric Hippogriff in the Harry Potter movies (see Photo 22).

By comparison, the elegant Sunbittern looks like nothing else: a large bird, half a metre long, with a long body compared with its height. The plumage is grey and rufous with attractive camouflage bars. The legs, slender neck and bill are of moderate length; the head is black with two white stripes from the bill to the back of the head. Unfortunately for us, the bird we admired simply stalked along a log and disappeared into the foliage, but had it felt threatened we may have been treated to a wonderful display, the wings opening out to appear huge, featuring big semi-circular chestnut patches edged with black on a mustard background, like scary eyes. The tail fans out to close the gap between the spread wings, so it resembles a huge fan or butterfly. It is close to being an even greater rarity – a species all on its own in a complete Order – but it is now generally agreed that the equally enigmatic Kagu of New Caledonia, which also forms its own complete Family, joins it as the only other member of the Order.

All this was pretty breath-taking, but then there were the Hoatzins! I regret to have to say that these splendid birds are, by any standards relating to most birds we know, seriously weird. You can't say 'it's a bit like a …', because whatever bird you were thinking of inserting for comparison, it isn't! It should be no surprise at all that their singularity doesn't end even at the Family level: they are unequivocally unique at the Order level. Only one other bird in the world (the Cuckoo-roller of Madagascar and the Comoros) can make that claim, though the ostriches, Kagu and Sunbittern come close for the different reasons discussed above. Hoatzins are always a highlight for me of any visit to the Amazon. I can't wait to see them, and I can't wait to show them to others. A Hoatzin is a substantial lump of a bird, 65 cm long and chook-like in general proportions, though in nothing else. It's not surprising that they look like nothing on Earth: astonishingly, their ancestors separated off from the rest of the land bird line 64 million years ago (i.e. around the time a mighty meteor took out all the dinosaurs except for the birds), and they have been flying solo ever since! (Prum *et al.* 2015) (see Photo 23).

The remarkable precocity of Hoatzin chicks, plus the unique possession of sharp claws on fingers two and three, has often induced speculation about direct links to dinosaurs, and specifically to Archaeopteryx, which is widely posited as the ancestral bird and which had three such wing claws. However, they have had plenty of time to come up with their own solutions to problems. Adult Hoatzins are big enough to be safe from most predators, but the chicks are not. Instead, even at just 3 days old they simply tumble into the water when threatened. Moreover, they dive and swim, kicking and moving their wings for propulsion. When they feel safe again, they scramble back into the trees, hauling themselves up with the claws, with neck, feet and beak. This is amazing behaviour for such young chicks.

Let me try to paint a Hoatzin for you (you could, of course, just go and look it up, but for now I would rather you kept reading). It is dark brown above with white bands across the wings and white streaks up the neck. There is a broad creamy band across the tip of the tail, and a rich chestnut panel in the wings (and on the flanks and lower belly, but you only see them when the wings are up). Then there is the head … a long spiky ginger crest that looks like a heavily gelled mohawk is perched above a bare blue face and dark red eyes. They are thus large and conspicuous, and are always along the edges of rivers and lakes, but often the first indication of their presence, as at Cocha Salvador, is a low cacophony of asthmatic huffing and hissing, with a miscellany of honks and hollow grunts. With this to guide the eye, we become aware of up to a dozen or more of the huge birds perched, albeit somewhat precariously, in groups on branches above the water. The problem is that they don't move much: they spend much of their day loafing.

Needless to say, there is a very good reason for this, and it is to be found in their diet. Hoatzins are most unusual among birds in being primarily leaf-eaters. This raises a very interesting question: vast numbers of mammals (and an even vaster biomass of them) are leaf specialists, relying on millions of tonnes of leaves (especially grass leaves), most of them eating nothing else from weaning to death. By contrast, hardly any birds do so; only some 3% of bird species, from only 14 Families of some 230 extant ones, eat leaves as a regular part of the diet. Why is this?

Birds eating their greens

A clue to these low numbers is in the additional observation that the majority of leaf-eating birds are larger terrestrial species – in addition to the Hoatzin, geese and ratites (emus, ostriches, rheas, etc.) are good examples. There are two problems with leaves as food. First, unless you're eating out for pleasure, the basic purpose of food is to obtain energy, and quite frankly leaves are a lousy source of energy. By weight, leaves only give half the return of fruit and a quarter that of insects and other wriggling goodies. Further, to get even that much return requires a veritable internal factory, run by a huge team of skilled and highly trained bacteria – no other creature (except for a few clever silverfish) can break down the complex sugars of cellulose, which form the fibrous mass of leaves. Grazing and browsing mammals have big fermentation chambers wherein the bacterial workforce does the job of digesting the cellulose. It is not a rapid process though – a Koala's gut can take up to 8 days to break down a eucalypt leaf, so a large caecum is required. In the Koala's case it is 2 m long and 10 cm in diameter: definitely not the sort of accoutrement you want to carry around with a strict baggage weight limit for your flight (in fact, as we will see, it effectively precludes a bird that uses this strategy from flying).

Foregut fermentation is a common mammal strategy, but the only confirmed avian example of it is in – yes – the Hoatzin. I guess that 64 million years of isolation is enough uninterrupted time to evolve a system that no other bird has managed. However, it takes 24 h to digest its salad, and at any given moment a Hoatzin is carrying something like 25% of its bodyweight in the form of digesting leaves in the foregut. There is another problem associated with this too: the foregut takes up so much room that the Hoatzin has had to largely do away with the sternum (the keel, as seen in a chook on a plate). This is the anchor point for the great flight muscles. As a consequence, a Hoatzin can barely fly, and in practice it mostly just scrambles through branches and glides.

Hoatzin chicks are fed by regurgitation on a soup of fermented foliage – not for them the protein that the young of most vegetarian birds are given to provide them with an early boost. There is one other apparent spin-off from this unusual digestive system too. In at least some parts of its range, the Hoatzin is regarded as smelling (like cow dung according to some reports) and tasting bad. In Guyana it is known as 'stinking pheasant', and 'stinkbird' and 'stinky turkey' are often cited as names elsewhere. It appears that this view is not universal, but, even in places where a revulsion to the bird is not reported, it seems to be eaten rarely, which would be unusual for such a large and easily procured bird. It may be that they are always unpalatable (or just smelly), but that additionally in some areas they eat toxic leaves. There are suggestions that Hoatzins may possess natural detoxifying chemicals, which are currently unidentified (Thomas and Bonan 2017).

But what about the other birds that do eat leaves? After all, 3% of 10 000 species is still some 300 species. To get a reward from your leafy lunch you don't actually have to go to all that fermentation trouble – by just grinding leaves up you can get to digest the cell contents without bothering about the nasty fibrous stuff, and that's exactly what all other leaf-eating birds do. However, this means that you are getting even less of the potential value of the salad, so you must eat more and have a bigger storage space, which means having a bigger body and/or a reduced flying capacity. There are very few small flying leaf-eaters. These include some of the saltators (South American tanagers – or possibly cardinals, but let's not get distracted now), the three South American plantcutters (cotingas) and the African mousebirds (small non-passerines). I've had the fortune to see all three groups in the wild, and although the mousebirds are certainly flutter-gliders, the other two are actually perfectly respectable flyers. This is a reminder, incidentally, that nature rarely feels constrained by the rules we write for her (they are usually our rules, rather than hers). The Rufous-tailed Plantcutter, found as far south as the stormy shores of the Strait of Magellan, is perhaps the smallest leaf-eating bird, relying almost solely on leaves in autumn and winter when fruits are scarce. The preference of all of these leaf-eaters is for the youngest, most tender shoots.

In Australia, the bowerbirds are among the few leaf-eaters, as any gardener in their territory will know. The most dependent on leaves is the Tooth-billed Bowerbird of the Wet Tropics of Queensland. It, like the plantcutters and the Satin Bowerbird, turns to leaves mostly in winter when there is a dearth of the preferred fruits.

The other problem with leaves is that so many of them are downright dangerous to eat. Plants have had hundreds of millions of years to build their defences against those billions of creatures – mostly insects – which would rudely chew them. There are more than 30 000 identified chemicals found in plants whose use is not apparently related to daily functioning: many of these are probably involved with defence. Many of the chemicals whose function we do know are pretty scary. Some wattles have cyanide precursors: break the leaf, bring the chemicals together, and you get a very nasty mouthful indeed. Milkweeds have cardiac glycosides, which are very unpleasant heart poisons. All in all there are very good reasons to look for other food sources, no matter how tempting and abundant leaves may be.

Fascinating as these very singular Cocha Salvador birds undoubtedly are, there was another species present which I was also very pleased to see.

Muscovy Ducks, and birds in service

In the shadows near the bank, two near-black ducks with a greenish sheen glided cautiously. The male was larger, with a white wing patch and big fleshy red growths at the top of his bill, which was banded black and yellow. Though I'd never before seen them in the wild, I was familiar enough with Muscovy Ducks, though the domestic ones I knew were mostly white. They were common domestic birds in South America when the conquistadors came with swords, cannon, bible and a rapaciously brutal certainty of their own rightness and superiority. Without written records (the South American cultures had sophisticated craftspeople and engineers, but seem not to have developed written language), we cannot know how long they had been commensals by then. However, given the long history of sophisticated farming practices and domestication by the Incas (and all the cultures that came before them), there is no reason to suppose that it was a recent event. There is some archaeological evidence for domestication by the Mochica people of southern Peru nearly 2000 years ago, though that wasn't necessarily the start of it either (Stahl 2005). Muscovies are still common as domestic birds in South America, especially in rural villages, but in the wild they are now uncommon throughout most of their extensive range and, very sensibly, mostly very wary of humans.

Muscovy Ducks arrived in Europe with returning Spanish looters in the 16th century. There is a great deal of confusion about the origin of their English name, and considerable intellectual gymnastics have been performed in an attempt to justify their connection with Moscow, but all such efforts seem doomed to failure. The reference was apparently to 'musky', in relation to a supposed characteristic of their flesh. The great 17th-century philosopher, scientist and theologian, the 'father of British natural history', John Ray, was quite unequivocal on the matter, and I think his proximity to the problem should be respected.

Regardless of this, these New World ducks were relative newcomers in a hemisphere where ducks had been part of daily human lives for at least 1500 years. The ancestor here and in Asia was the familiar 'Wild Duck' of Europe and Asia (and North America), the Mallard. However, the actual origin of duck domestication is mysterious, and seems not to have pre-dated the Romans in Europe at least (e.g. Albarella 2005). It is suggested that duck domestication in Asia preceded this, based it appears largely on the number of breeds present now, but direct evidence seems to be lacking. It would not have been a difficult process – ducklings from clutches of eggs taken from the wild and hatched under domestic hens would imprint on humans if they were present when the ducklings emerged.

Geese, however, seem to have been part of human lives for longer than that. Egyptian Geese were domesticated by the Old Kingdom of Egypt, which ended some 4300 years ago. Curiously, they were apparently (albeit inadvertently) 'liberated' by the Persians 2500 years ago when they conquered Egypt. Neither Persia nor subsequent Egyptian cultures renewed the relationship. European domestic geese derive from wild Greylag Geese: the oldest direct evidence is in the *Odyssey* (i.e. 2800 years ago), but there is a belief that the association is much older (e.g. Albarella 2005). Perhaps the most famous geese in our culture are the ones that supposedly guarded Rome. Actually it is not at all certain that they did, but the legend has it that a flock of domestic geese alerted the apparently dozy guards to the approach of the Gauls 2400 years ago; whether they were already sacred to the Temple of Juno, as reported, or were subsequently promoted, isn't clear. There is no doubt though that they are eminently suited to the role, being nosy, noisy, aggressive insomniacs. There are lots of stories of 'guard geese' since then, including in modern times in prisons in Colombia and Brazil, on South Vietnamese airbases in the 1950s, a whisky distillery in Scotland (until as recently as 2013), police stations in China and US military bases in Germany. Before we move on from European domestic geese, I do love the image of flocks of geese being 'shod' by walking them through tar and sand to protect their feet before walking them in large numbers and over considerable distances into town to be sold at markets (Carboneras 2017).

In China, Swan Geese were also domesticated millennia ago, though again the evidence for the timing is scant. One problem is that, where hearth bones are relied on as evidence, it is nearly impossible to distinguish between wild and domestic birds. Even in art, and where captive birds are depicted, it is very hard to say if they are domestic or newly captured wild stock.

Probably the most familiar poultry are domestic fowl (I actually think of them as 'chooks', but that's probably a bit too Australia-centric to use exclusively here, and I'm afraid that to me 'chicken' still implies a yellow baby chook). Although the Green Junglefowl of Java and nearby islands, and the Grey Junglefowl of India, were undoubtedly domesticated by people a long time ago, and have probably donated genes to modern farmyards and backyards, it seems almost certain that the Red Junglefowl is the major ancestor, brought into domesticity by the great civilisations of the Indus Valley in India 4000–5000 years ago. (These cultures also brought sheep, elephants, buffaloes, goats, cattle and camels into the fold.) From here they spread with traders, reaching north-western Europe, Egypt and China by 3500 years ago.

They were widespread throughout the south Pacific before Europeans got there. On many of these islands, runaways have reverted to the wild, and to the physical characteristics of their distant Indian ancestors. It was only to reach the Americas and (apparently) Australia that chooks required European assistance (McGowan and Bonan 2017).

There is archaeological evidence that Native North Americans have been domesticating Wild Turkeys for perhaps 2500 years. Cortés brought Aztec-domesticated Mexican turkeys back to Spain in 1519. The origin of the name 'turkey' is convoluted to the point of surrealism. According to one story, Helmeted Guineafowl were apparently taken from Africa to Turkey, perhaps via Portugal, in the 16th century. They had been kept as domestic birds in Egypt at least 4500 years ago, and in the eastern Mediterranean from around 2500 years ago, but disappeared from Europe with the fall of the Roman Empire. From Turkey they reached Europe and became known as Turkey-cock and -hen or simply Turkey (or Turky). However, it is also quite possible that 'turkey' was just a reference to 'somewhere foreign and distant' and that the country was never involved! Either way, when 'real' turkeys arrived a little later, the two birds became conflated (or were regarded as variants of the same bird) in the public mind. When this mental mist cleared, 'turkey' had become applied to the American birds, and 'guineafowl' to the west African birds (Porter 2017; Fraser and Gray 2013).

Venerable as is the history of these (conscripted) human companions, none of them are the longest standing of our feathered domestics. Rock Doves were first domesticated by grain farmers (who were involuntarily already feeding them), as mentioned in Mesopotamian records from more than 5000 years ago, though some believe that it could have happened as long ago as 10 000 years (Johnston and Janiga 1995). While we mostly think of domestic pigeons now in terms of racing and shows, at that stage, as with all the other domestic birds we've discussed, the original interest was in their culinary properties. However, pigeons have been used to carry messages since King Cyrus of Persia did so 2500 years ago, and continued to be used by the military well into the later 20th century (Westfahl 2015).

Their edibility has overwhelmingly been the primary focus of our interest in domesticating birds, but is not the only one. The imprisonment of small birds and parrots for human gratification is widespread and long-established – parrots have been caged for many centuries from India to Mexico. Today the most common cage birds in western culture are, to my great sadness, two arid-land Australian parrots: Budgerigars and Cockatiels. Both are essentially flock birds and (to borrow from a Jane Goodall quote about chimpanzees) one budgie is no budgie at all. The trade in small caged birds for household pleasure is enormous in South-East Asia; in Indonesia in particular a widespread and open illegal trade is driving species such as the Black-winged Myna to extinction (TRAFFIC 2015).

Atlantic Canaries were brought to Europe from the Canary Islands (named for dogs, not birds, but let's not get distracted) soon after 1400 and soon took over as the most-popular cage bird from European Goldfinches. They came down to earth abruptly when taken down coal mines to warn miners of gas infiltration (by dying) and they played this role in Britain until the late 20th century.

King Quail were reputedly used as hand warmers by Chinese emperors 5000 years ago. And, just to return to wetlands where this topic started, the continued use of Great Cormorants in China (for 2000 years) and Japan (for 1500 years) to fish for their controllers (albeit mostly now for tourists in Japan) is well known. Perhaps less-known, however, is the traditional use of Oriental Darters for the same purpose in South-East Asia and north-east India. Moreover, this practice (regarding cormorants) wasn't always exclusively an eastern pursuit. Members of the British royal court of James I delighted in the pastime for a while from the early 17th century. It ebbed and flowed in western Europe over subsequent centuries, but always as an upper-class pursuit, as opposed to its practical purposes in the east. There was a revival in England and France in the 19th century. Its derivation in Europe seems to have been independent of that in Asia, and the use of neck rings to prevent the birds from swallowing the fish seems not to have been adopted in Europe (Beike 2012).

Jerrabomberra Wetlands, Canberra

Just 3.5 km from the national parliament, and less than 5 km from the city centre, is an oasis of wetlands, creeks, native plantings and rough grassy paddocks where I have seen over 120 species of birds. Like Gum Swamp, it is a blend of natural and artificial. From a nearby hill, we can see it as a series of channels, seemingly artificial, across the open ground. In fact they are palaeochannels of the Molonglo River, gouged out in past big floods, and left behind when the river later returned to its own, or a new, channel. In pre-European times, they would have filled in wet years, then dried when El Niño returned. Now the flat shallow expanse of Kellys Swamp is kept watered in all but extreme drought conditions, as just downstream the Molonglo is backed up by Scrivener Dam to form Lake Burley Griffin, the centre point of the national capital. I have spent uncountable hours over the years – alone or in company, in good times and (occasionally) in bad – in the bird hides watching the ever-changing parade.

Preening

I have a vivid memory of a beautiful Yellow-billed Spoonbill just in front of one of the hides preening, running each wing feather in turn through its broad seemingly unsuited bill, making sure that each was in perfect working order (see Photo 24). Feathers are fundamental to a bird. Keeping them in perfect condition, with all the barbules tightly locked together to ensure that a warm layer of air is kept close to the skin and that the flight and tail feathers are working at maximum efficiency to ensure rapid lift and change of direction when required, can make the difference between life and death. (For details of feather structure and types, see pages 187–9.)

The spoonbill was meticulously 'rezipping' the hooks and loops along each barbule of each barb along each feather shaft so that each feather remains pristine. Many species – especially waterbirds, though not parrots or pigeons – use an oily secretion from a gland on the back to spread on the feathers. This cleans, assists with waterproofing, acts as protection against fungi and bacteria and maintains flexibility (which sounds like a commercial!). Many birds have independently evolved powder down: special feathers that grow constantly and never moult, whose tips crumble into a powder-like talc, which is used for cleaning and conditioning. It all sounds a bit pedestrian – and indeed a bird must spend a considerable time at it every day, often in the middle hours when feeding is slow – but I was mesmerised by the care and delicacy with which the spoonbill was performing its maintenance.

Further out on a mud flat, each of a group of Australian Pelicans was performing the same essential self-service. On the way back to the car later, I saw a Yellow-faced Honeyeater likewise assiduously engaged in the foliage.

Every day, in every part of the world where birds live, billions of birds are undertaking the same exhaustively repetitive, but utterly essential, activity.

Other memories of wetlands and rivers

Like bubbles in a churning stream, materialising on a dancing surface, catching the eye briefly before vanishing again into water and air, some briefly surfacing wetland memories:

- The Okavango Delta in Botswana is a remarkable place, and I was fortunate to visit it on a budget-price tour (of necessity, as much as preference), so had the privilege of entering it in a traditional mokoro (plural mekoro): a 3-m long dugout canoe with only 10 cm of freeboard. It is propelled by a poler, armed with a 3-m pole, forked at the end to prevent it sticking in the mud. We were enjoined to trust the poler and not to try to counterbalance when he leans the mokoro – that requires a lot of faith and self-discipline! Sometimes we travelled through alleys cut through the papyrus; sometimes we seemed to cut across country in very shallow water, at others in deep lagoons with the hippos and crocodiles … Birds leapt up whenever we emerged from an alley or turned a corner. And the constant soundtrack comprised the atmospheric wailing yelps of the African Fish Eagles. It was a powerful experience and seemed to mark the beginning of the next stage of my life – an excellent stage as it has turned out, and one that continues 14 years later.
- In Jerrabomberra again, gazing out of the hide early one morning across the mist rising from the waters of Kellys Swamp to the

imposing flag pole on Parliament House across the lake. Right in front of me, a long-billed cream and brown streaked wader is darting its bill into the mud at high speed: Latham's Snipe has flown from its breeding grounds in northern Japan to spend the southern winter in such wetlands. To have the bird in the same field of view as the National Parliament, which has ratified international treaties to protect it, seems very wonderful.

- I always seek out sewage ponds when I travel (and I am often surprised when the town tourist information office professes startled ignorance on the topic). I've seen some excellent birds at such places, but none can compare in my memory with Strandfontein in Cape Town, with Table Mountain rearing hugely behind, and the ponds stained pink with hundreds of flamingos. I loved it too that the sewage works at Paarl, north-east of Cape Town, has been delightfully renamed the Paarl Bird Sanctuary and comes with welcome signs, an information shelter, hides and picnic tables!

References

Albarella U (2005) Alternate fortunes? The role of domestic ducks and geese from Roman to Medieval times in Britain. In *Feathers, Grit and Symbolism: Birds and Humans in the Ancient Old and New Worlds. Proceedings of the 5th Meeting of the ICAZ Bird Working Group.* 26–28 August 2004, Munich. (Eds G Grupe and J Peters) pp. 249–258. Verlag Marie Leidorf, Rahden, Germany.

Alström P, Hooper DM, Liu Y, Olsson U, Mohan D, Gelang M, *et al.* (2014) Discovery of a relict lineage and monotypic Family of passerine birds. *Biology Letters* **10**(3), 20131067. doi:10.1098/rsbl.2013.1067

Andrewartha HG, Birch LC (1986) *The Ecological Web: More on the Distribution and Abundance of Animals.* University of Chicago Press, Chicago IL, USA.

Beike M (2012) The history of Cormorant fishing in Europe. *Vogelwelt* **133**, 1–21, <http://docplayer.net/5084656-The-history-of-cormorant-fishing-in-europe.html>.

Cadona W, Kattan G (2010) Territorial and reproductive behavior of the Torrent Duck (*Merganetta armata*) in the Central Andes of Colombia. *Ornitologia Colombiana* **9**, 38–47.

Carboneras C (2017) Ducks, Geese, Swans *(Anatidae)*. In *Handbook of the Birds of the World Alive.* (Eds J del Hoyo, A Elliott, J Sargatal, DA Christie, E de Juana). Lynx Edicions, Barcelona, Spain, <http://www.hbw.com/node/52210>.

Carboneras C, Kirwan GM (2017) Red-crested Pochard (*Netta rufina*). In *Handbook of the Birds of the World Alive.* (Eds J del Hoyo, A Elliott, J Sargatal, DA Christie and E de Juana). Lynx Edicions, Barcelona, Spain, <http://www.hbw.com/node/52899>.

Fraser I, Gray J (2013) *Australian Bird Names: A Complete Guide.* CSIRO Publishing, Melbourne.

Frith H (1967) *Waterfowl in Australia.* Angus and Robertson, Sydney.

Johnston RF, Janiga M (1995) *Feral Pigeons.* Oxford University Press, New York, USA.

McGowan PJK, Bonan A (2017) Pheasants, partridges, turkeys, grouse (*Phasianidae*). In *Handbook of the Birds of the World Alive.* (Eds J del Hoyo, A Elliott, J Sargatal, DA Christie and E de Juana). Lynx Edicions, Barcelona, Spain, <http://www.hbw.com/node/52221>.

Moyle RG, Chesser RT, Prum RO, Schikler P, Cracraft J (2006) Phylogeny and evolutionary history of Old World suboscine birds (Aves: Eurylaimides). *American Museum Novitates* **3544**, 1–22. doi:10.1206/0003-0082(2006) 3544[1:PAEHOO]2.0.CO;2

Porter WF (2017) Turkeys (Meleagrididae). In *Handbook of the Birds of the World Alive.* (Eds J del Hoyo, A Elliott, J Sargatal, DA Christie and E de Juana). Lynx Edicions, Barcelona, Spain, <http://www.hbw.com/node/52218>.

Prum RO, Berv JS, Dornburg A, Field DJ, Townsend JP, Lemmon EM, Lemmon AR (2015) A comprehensive phylogeny of birds (Aves) using targeted next-generation DNA sequencing. *Nature* **526**, 569–573. doi:10.1038/nature15697

Stahl PW (2005) An exploratory osteological study of the Muscovy Duck (*Cairina moschata*) (Aves: Anatidae) with implications for neotropical archaeology. *Journal of Archaeological Science* **32**(6), 915–929. doi:10.1016/j.jas.2005.01.009

Thomas BT, Bonan A (2017) Hoatzin (*Opisthocomidae*). In *Handbook of the Birds of the World Alive.* (Eds J del Hoyo, A Elliott, J Sargatal, DA Christie and E de Juana). Lynx Edicions, Barcelona, Spain, <http://www.hbw.com/node/52223>.

TRAFFIC (2015) Indonesia's illegal cage bird trade pushing Black-winged Mynas towards extinction. TRAFFIC, Cambridge, UK, <http://www.traffic.org/home/2015/8/13/indonesias-illegal-cage-bird-trade-pushing-black-winged-myna.html>.

Tyndale-Biscoe CH, Calaby JH, Davies SJJF (1995) Harold James Frith 1921–1982. *Historical Records of Australian Science* **10**(3), 247–263. doi:10.1071/HR9951030247

Westfahl G (2015) *A Day in a Working Life: 300 Trades and Professions Through History.* ABC-CLIO, Santa Barbara CA, USA.

6
Suburbia

Canberra: crafty cockies

Driving down a leafy inner Canberra street through a belt of parkland, I couldn't help but notice that most of the heavy transparent plastic streetlight covers were dangling below the exposed bulbs, holding on only by strips of rubber. It could have been interpreted as pretty shoddy construction, were it not for the presence a moment later of the culprits, cheerfully swinging from the damage or gnawing on the rubber seals until they gave way. Sulphur-crested Cockatoos are very handsome indeed: big porcelain-white cockies with a long yellow erectile crest, which have become very well-adapted urban birds in all Australian capital cities except Perth (they were introduced to the south-west in the 1930s but don't come much into the suburbs). Their beak grows constantly and needs to be kept in trim by gnawing – it may be that urban birds in particular, with a supply of soft fruits and backyard feeders stocked with other soft food, need to spend time in gnawing at random objects including wooden eaves, clothes pegs on lines and, of course, streetlights. They carry seed away in a crop – a thin extendable side-wall to the oesophagus – to digest in safety and at leisure, so have plenty of spare time on their claws. Moreover, they are highly intelligent birds and quite capable of investigating such objects just in case something edible is to be found inside. In recent times, urban Sulphur-crests in Sydney have learnt to open the heavy lids of big plastic wheelie bins to scavenge inside (ABC 2017).

Learning to live in town

This is just one of numerous examples of birds adapting to life in cities, the world's newest and fastest-growing habitat. A lot of work has gone into studying differences in calls of urban birds from those of their rural ancestors. Forced to compete against a loud background, urban-dwelling members of species such as Eurasian Blackbirds and Nightingales in Europe (Reinberger 2013) and Silvereyes in Australia (Potvin *et al.* 2011) sing more loudly than their country cousins. The Nightingales, however, toned it down at weekends when there was less traffic to compete with. Some species, including Great Tits and Eurasian Blackbirds, sing at higher pitches than they would out of town (Reinberger 2013). Some suggest that this is simply due to the stress of constantly yelling (try it yourself for a while), but a Dutch study (Halfwerk and Slabbekoorn 2009) showed that Great Tits switched to either higher or lower frequencies depending on what background noise they were exposed to – using higher to escape low frequency noise and vice versa. Artificial light is another factor that is entirely novel to birds moving into cities – or even just passing through them. Huge numbers of migratory birds in particular die annually by flying confusedly into lit buildings. One US Government study suggested an annual national figure of between 100 million and a billion such deaths (Kaufman 2011), but impacts can be much more subtle than that. The Max Planck Institute for Ornithology in Seewiesen in Germany is responsible for many such studies, including some of those already cited. In one of these, it was shown that some species – notably Eurasian Blackbirds, European Robins and Great Tits – begin to sing significantly earlier when artificial lights are present. Furthermore, another study demonstrated that, within European cities, male Eurasian Blue Tits living near lights were more successful suitors than those living further away. This may well simply be due to them starting to call earlier than their neighbours, but less easily explained is the fact that females of some species begin to lay eggs some days earlier if they live in the vicinity of a street light (Reinberger 2013). At the other end of the day, House Sparrows in New York and Bangkok forage later into the night, taking advantage of building lights, which attract insects (Donovan 2015). Around the world there must be millions of urban bird feeders dispensing tonnes of food a month to help birds through the tough times – or just to encourage their presence for the owner's pleasure. One might think that, as long as their owners stick to natural foods, this can only be a good thing, but not necessarily as it transpires. Suburban Florida Scrub Jays are getting prosperous on nuts and other vegetable-based goodies offered to them, to the point where their bodies tell them that because they're in such good condition they might as well start breeding. Accordingly, suburban jays have begun breeding weeks early, and laying more eggs. Sadly, their chicks need insect larvae as food, not nuts, and if they hatch before spring has properly begun – and they are doing so – they can starve (Anon 2006).

This perhaps leads us to the question as to which birds thrive in cities (it is clear that not all will – birds specialising in rainforests or open habitats, for instance, are unlikely to find their needs there). There are some interesting answers, and some debates. One widely cited earlier proposal (based on data) was that birds with larger brains did well – perhaps unsurprisingly, given the adaptations required (e.g. Maklakov *et al.* 2011). Other studies have firmly rejected this (e.g. Evans *et al.* 2011; Dale *et al.* 2015). The Evans study, across Britain, showed some evidence that eaters of plant parts (including seeds) and birds nesting above the ground were favoured, but a tendency towards being a habitat 'generalist' was a more important factor in urban success, again perhaps unsurprisingly. However, the Dale study, in Oslo, reminds us that sometimes the 'bleeding obvious' is where we should be looking: although preferred nest site and habitat preferences were important, the key issue in determining which birds were present in town was which ones were present in the surrounding countryside! Clearly there is a gradient from the greenless concrete canyons of the city centres through more or less leafy suburbs with gardens and parks, to the fringes where bushland is still present. In Australia, I think it is fair to say too that aggressiveness is a factor. Notably stroppy native birds such as Australian Magpies, Pied Currawongs, Sulphur-crested Cockatoos, Magpie-larks, Willie Wagtails and crows and ravens simply hold their ground against threats. Most of these – except for the fearlessly cranky little Willie Wagtail and medium-sized Magpie-lark – are large birds. Others, including natives such as Noisy Miners (a colonial honeyeater) and Rainbow Lorikeets, and exotics including Common Mynas and Spotted Doves, will actively attack and displace perceived competitors for feeding sources or breeding habitats. But, it is amazing what can turn up in suburbia, as any bird-aware suburbanite can attest. My previous address, where I spent a quarter of a century, was within a kilometre of the Canberra city centre – but significantly also within a short distance of the Australian National Botanic Gardens and the forested Black Mountain Nature Reserve, and just across the road from the Australian National University, which back then was full of green space (but in recent years has increasingly resembled a chronic building site). Over the years, my yard bird count – defined as birds seen in, over or from the yard – was close to 100. As the plantings grew, so did the list of visitors. One memorable winter, Fuscous and White-naped Honeyeaters regularly visited the saucepan of water hanging in front of my study window. (Fuscous are vanishingly rare garden birds here and White-napeds are not only uncommon in backyards, but are essentially migrants that only come to Canberra to breed in spring and summer.) As I watched one day, distracted from the computer screen, I suddenly felt as though I had strayed into an Attenborough documentary: a young Grey Butcherbird hurtled into the bathing flock of White-napeds, drowned one and flew off with it, the outraged flock in pursuit. None of those three species ever reappeared in the yard. In the last decade of my time there, a prolonged drought and catastrophic wildfires, on the edge of the urban area and even well into it, suppressed the numbers of smaller birds in particular, and the Noisy Miners muscled in, apparently displaced by the building works at the university over the road. After that, things were never the same. Nonetheless, there were still some remarkable moments. So much of birding seems to be chance, happening to be in the right place at the right time – though another way of putting that is that the more hours you put in, the luckier you're likely to be! One autumn morning I left my desk, slightly reluctantly, to investigate an avian ruckus outside – and saw my first Canberra Eastern Barn Owl blinking confusedly in a tree outside the back door, just before it panicked and was seen off the premises by the mob. Although Barn Owls are common enough across much of Australia, for some reason they rarely

Bird names in daily English

Let's go back to the Sulphur-crested Cockatoos, which have got bored with destroying the street lights and have gathered under the pin oaks in the adjacent parkland to crack open the iron-hard acorns. The power in their bills, with leverage assisted by a hinge where the top mandible meets the skull, is astonishing. Some birds are sitting in the trees above the flock, screeching abuse and flaring their crests as I approach. Whether they are really acting as sentinels is another question, though it seems quite possible, but the term 'cockatoo' certainly entered the Australian lexicon in this context, especially being used for someone who watched out for the police while illegal activities proceeded (particularly the coin-tossing gambling game known as two-up). According to the *Australian National Dictionary*, the term was in use as far back as 1827, recorded by the naval surgeon and social commentator Peter Cunningham. Cockatoo (or more usually just cockie) has yet another Australian meaning too: that of a farmer, based on the behaviour of some cockies, notably corellas, which forage for food by scratching in the dirt. There are subspecies of this particular cockie too, such as cow cockie or fruit cockie, but it was generally used disparagingly for a smallholder, though that has changed to a more neutral implication in recent times.

With the Canberra cockies among the acorns are a scattering of slightly smaller cockatoos: elegantly pink and grey Galahs, like a soft sunrise through morning clouds. These are relatively recent arrivals in south-eastern Australia, spreading from the north and west with crops and water points, and only arriving in Canberra in the 1950s. For reasons not entirely clear, their name has come to be applied to a foolish person, especially an exhibitionist one. Actually, once you've watched a flock of Galahs taking ages to settle down to roost, squabbling and whirling into the air over and again, or hanging upside down on the wires and harassing each other during the day, it's probably not too hard an idiom to fathom. (Much of the problem is that crop: the oesophageal sac that enables them to carry seed away to digest somewhere secure and comfortable.) Curiously 'galah' seems to have only come into written use in that sense in the 1930s, though doubtless it was in oral circulation before that.

We have form I'm afraid for using bird names as derogatory terms. 'Coot' tends to follow 'silly', for no overt reason. They're not bald though – when 'bald as a coot' was first applied, at least as far back as the 15th century, the original meaning of bald, or beld, or balled – that of having a white patch – was still in use; and, no, coots aren't white, but their bills are, at least in Europe and Australia. (The *Compact Oxford English Dictionary* – surely named with irony! – makes it clear that the origin of bald is somewhat mysterious, but favours this explanation.)

Even more shameful is our sneering appellation of 'booby' (i.e. a foolish fellow) to the nesting birds which were so trusting of sailors that they allowed themselves to be clubbed to death by the thousands. Turkey, goose, cuckoo are all terms for foolish or even slightly demented people. Sir William Hooker of Kew Gardens commented acerbically on 19th-century zoological artist William Swainson's ill-advised forays into Australian botany 'of which he is as ignorant as a goose'. A peacock is someone vain and pompous, though it seems to me that not all blokes who dress up and strut to impress females have feathers. Pigeons were also regarded as foolish and the word has a venerable history of describing an easy mark or sucker. A stool pigeon, now an informer, was originally someone used as 'a decoy to entice criminals into a trap' (World Wide Words 2001).

Drongo is used in Australia as a generally affectionate term of derision, implying a no-hoper, and its meaning too has changed with time. Drongos (and, in this case, the Spangled Drongo of northern Australia, New Guinea and associated islands) are in fact very sharp, speedy aerialists indeed, so presumably the owners of the 1920s Australian racehorse named it thus in the hope it would be as nippy. Despite the now reasonable assumption that the poor nag must have been totally hopeless, it wasn't, and even managed the odd gallant second. However, the horse never actually won, and so the original connotation was more of a trier who just fell short.

Rooster is an example of a recent political coinage that entered the lexicon, at least for a while. It was used by a prominent member of the Australian Labor Party, then in opposition, to describe a group of senior party members who he accused of undermining the party leader. It has since largely been forgotten, along with the person who coined it.

Other bird idioms are more benign. I suspect the term 'emu parade' may be sinking into history too, though anyone who grew up in Australia more than a couple of decades ago is likely to have been submitted to one, perhaps as an after-school group punishment. It comprised a straggling line of people moving across the ground to pick up rubbish, like a foraging flock of emus. In England, one who collects things, especially bright and shiny ones, is a magpie; in Australia the same person became a bowerbird. If you stretch your neck for a closer look, you're craning, or taking a gander. (To goose on the other hand refers to other, less savoury, farmyard behaviour, into which we need not delve too closely here.)

A few such terms are actually positive. 'Raven-haired' in the classics always seems to be followed by 'beauty'. 'Hawk-eyed' or 'eagle-eyed' are good ways of expressing exceptional visual acuity. Some people seem always to be able to swan along, looking relaxed and graceful. (I'm more apt to be duck-like, paddling furiously beneath the surface while trying to maintain a façade of calm competence above it.) Not that I've ever worn a dinner suit, but surely 'penguin suit' to describe one can only be positive! Less certain as to its flattery intent is the application of 'penguin' to nuns, though both refer to pristine black and white garb.

turn up in the ACT; there was a brief influx 4 years later, but that's all I've got to show for 37 years here. Even more astonishingly, 6 months later a Painted Buttonquail, a shy woodland specialist, turned up and spent a week in the yard – fortunately I have a photo to prove that one!

And on a summer evening I sat and watched enthralled until darkness hid from me the sky that was full of Pacific (or Fork-tailed) Swifts, from horizon to horizon. Just before flitting to another topic, however, I should mention an intriguing recent paper (Ives *et al.* 2016) that suggests that perhaps we *shouldn't* be too surprised when rarities pop up in our backyards. In a continental-scale study, every Australian town of more than 10 000 people (99 in total) was compared with distribution maps of every listed threatened Australian plant and animal species – 1643 of them. *Every single town has, or is likely to have, at least one threatened animal species.* Moreover, the study twinned each town with a 'dummy' town: an area of the same size randomly selected from the same bioregion. The real towns held considerably more threatened species than the dummy towns (i.e. equivalent rural areas). Without access to the appendices, I can't separate the birds out of the rest, but it provides nourishing food for thought.

Canberra again: the Great Koel debate

Some 20 years ago, I arrived home in inner suburban Canberra from my regular radio natural history slot, during which I had asked people to keep their ears open for a then unfamiliar call in town – the ringing rising 'COO-eee' of the Pacific Koel (as we now know it). I had confessed that I was yet to see one here, but when I got out of the car I immediately heard the clear blast of sound from the backyard. There, in the ornamental plum tree just over the back fence, was the magnificent big glossy black long-tailed cuckoo, his bulging eyes as big and red as the plums around him.

Those days seem long ago now: as I write, looking into a different backyard, I can hear at least two males yelling at each other, sometimes breaking into their alternative anthem, a manically ascending 'whoopa whoopa whoopa whoopa!'. It is an indivisible part of Canberra's summer soundtrack now, with large numbers arriving each spring and returning north to New Guinea and Indonesia in autumn. We have discovered in the process that a significant division of the world around here is now into those like me who revel in the wild calls cutting through the suburban hubbub of traffic (yes, even in the small hours), and those who definitely don't ...

In addition to controversy, the koels have been responsible for a considerable amount of entertainment here too in recent times. A newly elected opposition member of the Legislative Assembly was rapidly appointed as Shadow Environment Minister (it was then a pretty small assembly), and on behalf of some constituents demanded to know what the Environment Minister intended to do 'to eradicate or manage' this 'imported pest'. Oops!

Birds in a warming world

However, the real story is why this species has suddenly become a common migratory visitor where once it was very scarce. I am confident that the answer is largely to do with the warming of the planet; the koel is just a very audible part of a vast diaspora of species into areas previously too cool for them, towards the poles and up mountains, for instance. However, the cards also fell very propitiously for the koel. For one thing, Canberra gardens provide lots of fruit and berries that the big cuckoo relies on. On the coast these are readily available in rainforest pockets and other wet forests, as well as from the widespread and invasive exotic Camphor Laurel (*Cinnamomum camphora*), but in the woodlands, grasslands and dry forests around Canberra they are in short supply. So, having ventured south and inland they found a bounty awaiting them here. But something else was of great value to them too – the abundant breeding Red Wattlebirds, big strong honeyeaters quite capable of raising demanding koel chicks.

But this too contains an interesting story. During the 20th century, the wattlebirds moved north along the east coast (perhaps assisted by orchards, though this isn't clear) until they reached Sydney (Blakers *et al.* 1984) – where the koels met them. Now the koels had hitherto laid their eggs in the nest of another big honeyeater, the Noisy Friarbird, plus Australasian Figbirds and Magpie-larks, all of which probably have at least some evolved defences. However, the Red Wattlebirds, with just the right-sized nest, knew nothing of big cuckoos taking over their brood and were totally defenceless. The koels only cottoned onto this possibility in the 1970s and within a decade the unfortunate wattlebirds had become the main host of the koel in Sydney (Brooker and Brooker 1989). And it seems they still haven't learned. Australian National University Ph.D. student Virginia Abernathy experimented with artificial eggs in Red Wattlebird nests. Unlike most cuckoo hosts, the wattlebirds had no idea that they could or should eject the eggs, so are totally vulnerable (Professor Naomi Langmore, *pers. comm.*). A koel's dream!

But the ultimate driver of all this was the warming world. In early 2017, a pair of Tawny Grassbirds, a skulking but fairly vocal non-migratory species from considerably north of here, appeared in a major Canberra suburban wetland; this followed records from the coast near Melbourne, well south of here. I recently heard the unmistakably frenetic and slightly wheezy honking trumpet call of a Channel-billed Cuckoo go past outside, but was unable to see it – I wasn't too worried because I am sure that it is only a matter of time before this one too becomes a regular visitor, with Canberra records increasing by the year. Its annual migratory cycle is very similar to that of the koel.

All over the world, evidence is piling up of organisms moving their ranges – fish, invertebrates, frogs, mammals and even plants (not individual plants, of course, but whole communities are moving up mountains, from the Andes to the French alps). And if you're moving up a mountain or towards a pole, there comes a point when there is nowhere else to go.

It is not just movement, though. Behaviour is also changing, especially in phenology – the annual cycles that determine fundamental behaviour such as breeding and behaviour. As far back as 2003, a wide-ranging review in the prestigious journal *Nature* revealed 'significant mean advancement of spring events' by 2.3 days per decade (Parmesan and Yohe 2003). Four years later, the Intergovernmental Panel on Climate Change's 2007 report revealed that the arrival of spring had been advanced by up to 5.2 days per decade over the past 30 years. Examples cited ranged from first and last appearance of leaves on Gingkos in Japan, to butterfly emergence in Britain, to bird migration in Australia (IPCC 2007, p. 99). A more recent comprehensive Australian review of 89 studies of 347 plant and animal species concluded that the spring migration departure of birds moved forward by 2.2 days per decade (less than for the Northern Hemisphere, where it was 3.7 days per decade) (Chambers *et al.* 2013).

One of the problems with all this is that, naturally enough, each species has a slightly different response to the changes, so that finely tuned systems are no longer functioning as they evolved to do. Bird chicks hatch before or after their key caterpillar food supply is available, migratory birds are arriving before or after the flowers they pollinate are open.

But wait, there's more ... In recent times, a third general response has been suggested, and demonstrated. Although there are always multiple factors acting on the life and evolution of any given organism, we know that, in general, body size of a given species is likely to be smaller in populations further from the poles (i.e. in warmer climes). This is known as Bergmann's Rule and the basis of it is that a smaller object (be it bird or ball or human baby) has a proportionately greater surface area than a larger one, and thus loses heat faster. If you live in a cold place, it makes sense to be larger to retain heat better. We know that this is the case for populations of the same species at different latitudes, but what about the same species at the same latitude as climate changes (i.e. the environment gets steadily warmer)? A treasure trove of such data is held in museum specimens throughout the world.

Janet Gardner of the Australian National University, and colleagues, measured 517 museum skins of eight Australian insect-eating birds, collected over 130 years from 1869 to 2001. Six of the species (Variegated Fairy-wren, Yellow-rumped Thornbill, Hooded Robin, White-browed Babbler, Brown Treecreeper and Jacky Winter) showed a decrease in size since 1950, with four of them being statistically significant. The overall impact for those four bird species is that individuals living now at the latitude of Canberra are the size that members of their species were pre-1950 at the latitude of Brisbane (i.e. 7 degrees of latitude) (Gardner *et al.* 2009). This I find very striking. Nor is it simply academic – a change in size of even just 4% (as measured in wing lengths by the study) can affect what a bird eats, and thus what it is competing with and must further adapt to.

Only 10 years ago this particular response to climate change was only being guessed at, and there will be more surprises to come. An example, whose detailed explanation has yet to emerge, has been revealed by two European owl species, both of which come in two colour forms. In the case of the Eurasian Scops Owl, its colour forms are dark-reddish and pale-reddish (and intermediates). Italian museum studies showed that the proportion of dark-red forms increased significantly over the last century. Some of that was due to unknown causes (perhaps an increase in Italian forests over that time, where being darker could be advantageous, suggest the authors) but the rest is apparently down to climate change. At this stage, the best explanation is that the gene for dark-red is linked to one that confers an advantage in a warmer world, but so far we can only speculate (Galeotti *et al.* 2009).

The same is true of the similar case of the Tawny Owl, which has a grey and a brown morph, in a simple genetic system where brown is dominant to grey. Historically, browns have formed a minority of the population and, for reasons uncertain, are less viable in very snowy winters, such as have historically been the norm in Finland where the study was conducted. In recent times, however, winter snow cover has decreased and the proportion of brown owls has steadily increased (Karell *et al.* 2011). The reason for grey being better than brown in the snow, but not otherwise, is not clear, and again may well not be directly related to the colours – it could be another character that is driving the success, which happens to be genetically linked to brownness. However, both owl stories are clear cases of climate change directly driving evolution.

But can we say that a warming world is bad for birds? Unfortunately, in at least some cases, and especially in already hot dry situations, we certainly can. In a discussion of strategies for keeping body temperature to a safe level, while avoiding dehydration (pages 17–20), we saw that, as both frequency and intensity of extreme air temperature events rise, both heat shedding by radiation and convection from the bill, and by evaporation through panting, may become inefficient, and even hazardous. The crucial window of opportunity for keeping cool becomes a dangerously narrowing crack. Albright *et al.* (2017), in their study of passerines in the deserts of south-western USA (see page 19), concluded that, by the end of the 21st century, the smaller species they looked at will be exposed to at least four times the number of potentially fatal temperature events as they are now.

Gardner *et al.* (2016) analysed the data from long-term banding studies in semi-arid woodland in New South Wales, looking at the effect of long-term exposure to higher, but non-lethal, temperatures on White-plumed Honeyeaters (a very common woodland and semi-arid land bird). By comparing the measurements against climatic data, they found that repeated exposure to temperatures in excess of 35°C led to a 3% loss of body mass per day of exposure, especially in the absence of rain in the previous 30 days. This may be either directly due to dehydration, or because foraging is harder in such conditions, or more likely a combination of both stresses. During the 26 years of the study data, the temperature increased on average by 0.06° per year, the number of days over 35° increased and rainfall decreased. In that same time, summer survival rates (as measured by recaptures in the following winter) fell disproportionately among smaller birds, which tended to be females. So, although smaller individuals of a species seem to be advantaged in a warming world in general, in more extreme situations this is not the case. It comes back to the relatively larger surface area of a small bird, which applies not only to the body surface but to internal surfaces such as the mouth and upper respiratory tract through which birds seek to lose heat via evaporation by panting. Smaller birds in such conditions are constantly water stressed, seek to reduce stress by sitting still, so are not foraging enough, lose more weight and thus are in a deadly spiral.

Overall, the news is bad for birds living in already physiologically stressful arid lands if they are warming and drying even more.

Melbourne Cricket Ground: a groggy gull

January 2015, and a night cricket match was underway (well OK, it shouldn't really be called a cricket match, since the entire contest was only scheduled to take 40 overs, but it did have some similarities). A ball was swiped through the on-side and was clearly en route to the boundary, but instead was seriously impeded by a Silver Gull loafing without due care and attention. The bird understandably collapsed into a bedraggled heap and was carried to the boundary by a bemused fielder (who mimed to someone on the sideline that a coup de grace per large firearm could be required, but this is Australia, so who did he think would have such a weapon at the cricket?!).

The Silver Gull is one of the world's smaller gulls, an abundant bird around the coast of Australia and well inland, even breeding on the desert salt lakes in the very centre of the continent on the rare occasions that they flood. In this situation, they prey on the eggs and chicks of other breeding birds, to the extent that control programs have been instituted on occasions to protect the rarer species. Silver Gulls have adapted with enthusiasm to human habitation, gathering in flocks at rubbish dumps and anywhere that food scraps are available. Opening a parcel of fish and chips at the beach can be a hazardous undertaking, as one is inevitably and instantly surrounded by dozens of shrilly demanding gulls. Sadly, I suspect that many people believe that chips are in fact their natural food. And the gulls roost on open areas for safety, including under lights in order to see approaching predators – this predilection certainly includes cricket grounds and, as a result, they may be one of the most incidentally televised birds in the world.

Back at the G, as the ground is universally known here, the bird eventually staggered groggily to its feet to the astonished delight of the crowd, which rose in a standing ovation. It stood gasping for a while – then to the acclaim of the world (this story really was reported across much of the globe) flew unsteadily right back to where it had been previously stationed. (Some outlets claimed the bird was attacking its former rescuer – presumably for his shooting gesture – but a quick scrutiny of the footage makes it clear that this was not part of its plan.)

When birds collide

Some years ago I was given a delightful 1953 book called *Murray Walkabout* by Archer Russell. In it he recounts an apparently very tall story wherein he describes a Sacred Kingfisher killing a Grey Teal. 'Straight as a javelin it flew, straight into the middle of the covey it flashed. As the kingfisher flew among them, the teal rose in their flight and then sped on – all but one. That one first swooped and then fell headlong into the water ... The kingfisher had struck the teal on the head, apparently piercing the brain and killing the bird instantly'

(Russell 1953, p. 27). He makes it that clear he regarded it as an intentional attack. It has worried me since I read it, because it seems to be such far-fetched nonsense in the context of an otherwise generally sensible and credible book. However, it recently occurred to me that perhaps he really had simply observed a remarkably unlikely accidental collision and misinterpreted it.

And while this event, if it happened thus, is anything but normal, such unfortunate conjunctions of time and place involving birds happen all the time, not least on roads. I've had the misfortune to hit more birds than I want to think about in the course of driving hundreds of thousands of kilometres around Australia, but one bizarre event sticks in my mind. I was driving early one morning through the vast wild expanse of the inland Pilliga forests: 100 km of mesmerisingly beautiful ironbark and box eucalypts and cypress pine. I winced as a Laughing Kookaburra flashed across in front of the car, relaxed, then watched in horrified fascination as it swung high into the air, turned back and homed in with what seemed like determined inevitability on the windscreen. Out of the whole vast grey-green expanse of the Pilliga, that windscreen was a pretty tiny dot at that precise moment in time, but it and bird coincided, to the disbenefit of both. And why not? I know of smaller objects and smaller birds doing so, no matter how improbably.

In the Marylebone Cricket Club museum at Lords in London is a House Sparrow mounted on the cricket ball that prematurely ended its career in 1936, when it flew across the pitch during an MCC *v.* Cambridge University match, a moment after Jehangir Khan had released the ball. Tom Pierce, the batsman, was left waiting for the ball, which never arrived. (Jehangir's son, Majid, went on to captain Pakistan, but never matched his father's achievement.)

I don't want you to think that I'm cricket-obsessed, but I know of at least one other bird that died in mid-pitch, this one an even nippier Welcome Swallow at the Adelaide Oval in 1969. Greg Chappell, then a young bloke, was bowling to John Inverarity, at that stage still representing Western Australia. The ball suddenly dipped alarmingly and bowled the perplexed Inverarity. When he was half-way off the ground the honourable South Australian fielders called the umpire's attention to the dead bird lying some distance behind the wicket, and Inverarity got called back. It beats me how no-one saw the bird being hit, but there you go. I also wonder if the fielders would have mentioned it these days.

I am reminded too of another ex-sparrow whose much more recent demise was premeditated. It was shot in 2005 in the Netherlands – presumably after a fair trial and all appeals had been exhausted – for slipping through an open window and pre-empting the felling of 23 000 dominos, just before the lucrative television coverage to 11 countries began. But that is irrelevant to this story.

Inspired by these unlikely events, I went looking for others, and in the process found myself straying into the strange parallel universe of YouTube. (And in case you think I am being a little unkind to that monumental institution, consider that some of the following incidents, involving the sudden violent death of unsuspecting birds, are filed under Comedy.) A left-handed baseball pitcher hurls the ball – and a bird taking a short cut (either gull or pigeon I think) explodes in a cloud of feathers. The snout of one of a group of performing dolphins leaping out of a pool clips a passing gull, which cartwheels into the water. In a doubles tennis match, a small passerine (possibly a swallow) swoops across the net: ball and bird stop dead in their respective tracks. In Buenos Aires, a shot on goal by a soccer player results in a dead pigeon.

Birdies inevitably feature in golf. A woman – possibly a touch more amateur than some of the previously mentioned (human) participants – tees off and, unfortunately for the gull foraging some 15 m ahead, the golf ball doesn't get above ground level. On the other hand, professional golfer Tripp Isenhour was charged and sentenced in 2007 for deliberately killing a Red-shouldered Hawk with a golf ball, to punish it for making a noise and interrupting something vital, but unspecified, which he was being filmed doing.

At the other end of the scale, a space shuttle rocket, weighing apparently some 2000 tonnes, clips an unidentifiable (but presumably large) bird with the cone. It appears to not be moving very fast, but the inertia is irresistible. And perhaps with that it is time to change the direction of this narrative …

Canberra yet again: hybrid parrots

In the leafy backyard I mentioned earlier, I put out seed for parrots a couple of times a week. Over the course of a year I could expect up to seven parrot species to drop in, while a couple more just passed over. One pair of visitors I found particularly intriguing. One was a Crimson Rosella, the commonest parrot visitor; the other, however, was a young bird, and clearly a hybrid with an Eastern Rosella. The Australian Capital Territory is an interesting area biologically, in that here three major habitat types come together. The high subalpine woodlands and alpine heaths of the Australian Alps reach their northern extent, the wet montane forests of the coastal hinterland are close to their most inland limits and the great grassy woodlands of the western slopes of the Great Dividing Range sweep down from the north and west. Generally, Eastern Rosellas are woodland birds and Crimsons are dwellers of the mountain forests, so the situation in Canberra where they live side by side is actually an unusual (though not unique) one. Before the coming of a city, they probably very rarely saw each other, with the Easterns staying out on the plains and the Crimsons in the mountain and hill forests, but gardens and parks provide elements of both those habitats so they have become neighbours, of us and of each other.

Becoming a species: it doesn't happen overnight!

For those imbued with the axiom that species represent populations that are unable to interbreed, the existence of such a hybrid is a challenge – but it ought not to be. Speciation, whereby isolated populations of a species develop in different directions from those taken by their now distant relatives, to the point that they can't interbreed if and when they do come back into contact, is best thought of as a process rather than an event. Let's call the ancestor of the closely related Flame and Scarlet Robins, relatively common and familiar birds in south-eastern Australia, a Flarlet Robin. When something intervened – most likely changing climate associated with the glaciation cycle of the current ice age – pushing different Flarlet Robin populations into different habitats, it was just the start of a steady journey towards eventual species-hood. Each population was facing different challenges, different climates, plants, prey and predators. Adapting to all these, plus more or less random genetic variation, drove changes that were ultimately critical. We can imagine the start of the journey, and we can see the end-point (at least so far), where Flame and Scarlet Robins can and do live side by side but never interbreed. However, there was not a moment when a Flarlet Robin went to sleep and awoke as a Scarlet Robin, nor did a Flarlet Robin lay eggs which hatched into Flame Robin chicks.

We don't have a problem with such a concept in other day-to-day processes. If we choose to grow our hair long (and I refuse to believe that anyone of my generation didn't at some stage – actually no, to be fair, I don't ever recall a male engineering student doing so), we don't expect suddenly to be able to say 'aha, now it's stopped being short and is long!'. If I were to let a glass of beer sit untouched – well OK, that's pretty implausible, let's make it a sparkling water – I wouldn't be able to sit and observe it, saying 'sparkling … sparkling … sparkling … *now*, it's flat!'. We define the beginning and end points and accept that there's a continuum in between them.

But in other matters we have an obsession with being able to pigeonhole and label things precisely, which makes it very hard for us to understand the process of evolution. We are lucky in that the intermediate stages of the journeys from Flarlet Robin to Scarlet and Flame Robins are all dead and lost to us. If we could see the long, long chains of individuals, each externally indistinguishable from its parents and its chicks, our frazzled minds might be forced to concede that Scarlet and Flame Robins can't, so don't, really exist! (For a much more detailed and erudite discussion of this, I don't think we can go past Richard Dawkins' story *The Salamander's Tale* in his marvellous *The Ancestor's Tale*; Dawkins 2005).

Life is like an old-fashioned movie, comprising vast numbers of individual snapshots strung together; it is by merest chance that we are now in *this* frame, and not any of the thousands of other possible ones that precede and follow this one. And in this frame Flame and Scarlet Robins were separated for long enough that the journey to species-hood is completed, and they can never produce joint young again. They don't recognise each other as potential mates, and their internal chemistry would almost certainly prevent successful fertilisation even if such interaction were to occur.

For the rosellas, however, in this frame that journey is not quite so advanced: they don't generally interact and, when they do, as in Canberra, they still nearly always maintain their individuality. Occasionally, however, something goes wrong: perhaps they are young inexperienced birds, maybe they lost their parents early, perhaps some had even escaped from aviaries where they never learnt how to be an Eastern or Crimson Rosella. In such unions, we would expect the offspring to be infertile, but again it is a matter of timing in the speciation journey, and I have seen some circumstantial evidence that my backyard hybrid may have bred, but the line apparently eventually died out and there is no evidence of a spreading pool of rosella hybrids in Canberra.

But what about those birds that in this frame are not nearly so far down the speciation track? Australian Magpies (named for their superficial similarity to the Northern Hemisphere magpies, which are crows, to which our magpies are not related) are found right across the continent. There are eight Australian subspecies recognised (plus one in New Guinea), which fall into three well-defined groups, which used to be regarded as separate species. The White-backed Magpie of near-coastal south-eastern Australia and the Black-backed of most of the rest of the country are readily distinguished, as the names imply. The male of the Western Magpie of the south-west of Western Australia is identical to the White-backed, while the female has an attractive black back with white-edged feathers.

It has been suggested that in the glaciation before last (i.e. around 100 000 years ago) Black-backed Magpies entered Tasmania when the floor of Bass Strait was exposed as the Bassian Plain, and were trapped there when the ice caps melted and the seas rose again. In the 70 000 or so years until the next glaciation and drop in sea level, a mutation for a white back appeared and became ubiquitous in Tasmania; it seems that the genetic difference between the two forms is very minor indeed. At the time of the most recent glaciation (roughly 20 000–25 000 years ago), these new White-backed Magpies invaded the mainland and began to push north and west along the coast, rolling back the Black-backs (Burton and Martin 1976). This is not confirmed: Mitterdorfer, working with a limited number of genetic markers, suggested that the mainland White-backs are actually more closely related to the Black-backs than they are to the Tasmanian birds, which implies that a different, older event separated mainland ancestors of Black- and White-backed Magpies (Mitterdorfer 1996). The Tasmanian origin appeals (offering an easily envisaged narrative), but that doesn't make it right and we'll have to wait a bit longer for the definitive answer.

All magpie subspecies readily interbreed where they overlap. Canberra happens to be in the zone of hybridisation of Black-backed and White-backed Magpies, so that, although Black-backed is the dominant form in most of our yards, White-backeds are not uncommon. Where they interbreed, their chicks exhibit a full range of back colour from all white to all black (i.e. different proportions of each; they don't blend to become grey). This hybridisation zone is roughly 120 km wide, and on the east coast it is moving steadily north. In the past 30 years, I've seen it move from the far south of the Australian Capital Territory north to near Canberra. In the light of our earlier discussion about climate change, it might be beguiling to attribute this movement too to global warming, but there is no evidence that there is any climatic advantage to being either black-backed or white-backed. It seems that the more prosaic explanation is the correct one: White-backs are simply a little larger, and are slowly pushing the Black-backs north.

An intriguing aspect of this hybrid zone is its stability, rolling slowly north but not expanding into an ever-widening blur of hybrids. It seems that magpie behaviour is the explanation for this: non-migrants, they are strongly territorial breeders that cling fiercely to their territory, and even birds that haven't yet attained territorial status just hang around on the fringes awaiting their chances.

Given the existence of eight abutting subspecies, clearly magpie populations have been separated many times in the past, altered to some extent (and in the case of back colour, quite strikingly) but have always come back together before enough time had elapsed to allow them to change to the point of sexual isolation. It is unclear how much human alteration to the environment has facilitated this coming together of the races, but clearing of forest for farmland has undoubtedly benefited the species and its numbers and range have greatly expanded, especially in the east, so it seems likely to have been relevant to at least some extent.

There is another familiar Australian species, however, where our role in interrupting evolution seems more clear cut. The bird I grew up calling a Spur-winged Plover, but now formally known as the Masked Lapwing, is a large stroppy plover with a brain-piercing war cry, which lives in open country throughout northern and eastern Australia. There are two clearly defined subspecies, which, like the magpie subspecies, used to be (and sometimes still are) regarded as full species. In the south, the formerly named Spur-winged Plover has relatively small yellow facial wattles, extensive black on its crown and conspicuous and intimidating black-tipped yellow bony spurs pointing forward from the angle of the wing. It uses these spurs without compunction while swooping and shrieking at anything that dares come within the declared no-go zone surrounding its nest on the ground. Given that this nest can be on a school oval, playing field or grassy roadside or centre strip, this regularly creates anxious stand-offs!

The tropical Masked Lapwing (or Plover) has enormous yellow wattles framing its face, much less black on the head and less obvious wing spurs. At the time of European settlement, it seems that the populations were separated (at least along the coast, though there is some doubt about the situation in central Australia) and heading for separate species status. However, the clearing of forests for pasture, airstrips, roadside verges and ovals created plover Nirvana and the populations spread north and south until they reunited in the vicinity of Townsville in north Queensland. Here they found that their differences were still not significant enough and they interbred, forming a broad hybrid zone, and that particular line of evolution was snipped off.

Still in Canberra: pardalotes

An interesting thought has really only just struck me; I've been fortunate enough to have enjoyed birds in quite a few, mostly Southern Hemisphere, countries, but other than chance encounters I've done little birding in cities other than my own. I guess I just arrive and head for the bush for the most part! Still, the birds and their stories that I've introduced in this chapter could as easily have been in several other Australian cities, so I hope you've not been too put off by my apparent partiality.

For some weeks in spring the compost heap had to be quarantined while a pair of exquisite tiny birds took it over, burrowing deep into the soil near the bottom.

Spotted Pardalotes are close to being Australia's smallest birds, barely 10 cm long and weighing only 10–12 g, but their presence is powerful well beyond their physical size. The male has a black head, wings and tail generously spangled with clearest white spots, a broad white eyebrow, buff-scalloped back, red rump and bright yellow throat, breast and vent, while the female is somewhat more muted in colour. The volume of sound they produce is startling for such a scrap of bird – a clear ringingly repeated 'dee-DEE-dee', delivered in measured tones, which seems part of the summer air. They are common urban birds, as long as there are eucalypts to feed on: they specialise in gleaning leaves for small insects or spiders. In particular, they focus on lerps, sap-sucking bugs that use the surplus sugar they absorb, while trying to get enough nitrogen, to build little protective shelters over themselves. On a still hot day, the pattering of the lerp covers onto the dry forest floor can be surprisingly audible as pardalotes nip them off and discard them to attack the insect.

The pardalotes form their own endemic Family of just four species, in one genus. They are very similar in form: all minute with stubby bills and very short tails. The first edition of my favourite field guide described them as like 'flying beetles' (Pizzey and Doyle 1980). (Sadly, later editions had no room for some of those elegant perceptions, which were a hallmark of the great Graham Pizzey, who is unfortunately no longer with us.) Spotted and Striated Pardalotes are very familiar to city dwellers along the east, south and south-west coasts, though Striateds are also found across most of the country. ('Striated Pardalote' is a delicious oxymoron, incidentally, because 'pardalote' is directly from the genus name *Pardalotus*, meaning 'leopard-spotted'. Worse still, the 'Striated' refers to faint crown-streaks that are only present in southern Australian birds.) The plainer Red-browed Pardalote also has a huge range in the dry inland as far as the north coast, while the even more modestly attired Forty-spotted Pardalote is a Tasmanian endemic. All nest in hollows, either ground-excavated or in tree cavities.

In the compost heap, the female Spotted Pardalote seemed to be doing most of the initial work, breaking the surface first with her bill and then digging with her feet (not that I could see what she was doing once she got below the surface, of course, just the spurts of soil erupting from the entrance), but both birds took over after that. They would have continued horizontally for anything from 25 cm to a metre, digging a circular hole 4–5 cm in diameter, which eventually expanded into a chamber up to 10 cm across which they line with fibrous eucalypt bark and dry grass. Like other pardalotes, they lay three or four white eggs – no need for camouflage in a burrow – that they incubate for about 3 weeks, after which the chicks take another 3 weeks or so to fledge. This is in contrast to those of other small foliage-gleaners, which build open cup nests, lay fewer eggs and get them into the world more quickly (Woinarski and Bonan 2017). It seems that the burrows allow a more leisurely approach, which implies less stress on the parents.

Spotted Pardalotes will sometimes use building cavities or even hanging plant pots, but mostly concentrate on excavated burrows, as do Red-broweds. On the other hand, Forty-spotteds usually use tree hollows, and Striated Pardalotes use both fairly equally (though curiously not in Canberra, where they mostly stick to trees).

Birds in burrows

A burrow makes great real estate! Even in blazing summer, it remains relatively cool and moist and in winter it stays much cosier than the blizzard above. It can provide pretty fair fire protection too. (And no, I have no vested interests in the burrow marketing industry!) A burrow also greatly limits the predators that can get at you; however, unless you provide extra exits, it can be pretty dodgy when an enemy does come calling. The pardalotes certainly know that, exploding out of the hole with a leap and opening their wings when already airborne. Their re-entry takes the form of a headlong dive. Roaming cats will wait in ambush by the burrow and I did my bit to help by putting a wide-meshed cage over the hole, which gave the birds a bit more of an edge.

When we think of burrowing animals we are most likely to think of mammals, and with very good reason: many mammal species, from at least 30 Families that I can think of, live or breed in burrows. Many reptiles do too, not to mention invertebrates because they are just too numerous to cover meaningfully! But birds don't spring readily to mind in this context. Why don't more birds use burrows? For a start, delicate bones and a bill are not generally the best digging equipment. Then there are the wings: a bird's best escape mechanism is not going to work underground. Moreover, those same wings mean that a bird can much more readily escape the heat, aridity, cold, fire or predator that surface-bound animals use a burrow to avoid, so they have less need of one.

Despite all these reasons, many birds do at least breed in burrows, though, as far as I know, only the little Burrowing Owl of the Americas lives in one all year round. However, when I counted, I was surprised to find that there as many bird Families containing burrow nesters as there are mammal Families, though I also note that there are ~50% more Families of birds (roughly 235–240) than there are of mammals (155–160) (see Photo 25).

I can think of some 30 bird species that nest in burrows in Australia. Some – including some of our rarest tropical parrots and the apparently extinct Paradise Parrot, plus eight kingfisher species – dig into termite mounds, both arboreal and terrestrial. This comes with an added bonus for the eggs and chicks, because the hosts undertake a certain amount of temperature control. The downside is that termite mounds are *hard*. It is painful to read accounts of kingfishers flying flat out and bill-first into a rock-hard mound to chip away at the hard shell (e.g. Hollands 1999). Fatalities have even been recorded in the attempts (Woodall 2017). Once inside, female Buff-breasted Paradise-kingfishers use their impressive tail to sweep dirt out (Woodall 2017).

Presumably, the hooked bills of the parrots are more suited to the task. Golden-shouldered Parrot females do most of the work, digging with the bill and scraping dirt away with the feet. They have to get their timing right, however: if they start too early in the wet season, the termites will simply seal the holes again, or even cement the eggs to the floor, presumably as part of nest management (Garnett and Crowley 2017).

If digging into soil, one can select a more favourable substrate, though, if it is too easy to dig, the hole is also likely to collapse. Rainbow Bee-eaters dig in a flurry of activity, dislodging soil with the bill while balancing on wings and feet, then hurling it out behind with pedalling feet, balanced now on a triangle of bill and wings. Their burrow, which can be 1.5 m deep, progresses at ~8 cm a day. Webbed feet obviously make good shovels, because a pair of Wedge-tailed Shearwaters can apparently knock up a 15 cm diameter, 1.5 m long burrow, plus a 30 cm chamber in the end, in a mere 6 nights work (though I confess that I can't now actually find where I read that).

Remarkably, on Rottnest Island off Perth, Western Australia, it has been calculated that each hectare of a Wedge-tailed Shearwater colony comprised more than 5 km of tunnels, involving the displacement of 210 tonnes of soil (Bancroft *et al.* 2008)! Other burrowing birds are pretty assiduous in their excavations too: one of the most impressive is the Burrowing Parrot (or Patagonian Conure), mostly of Argentina. Their burrows can be 3 m long, zig-zagging into cliff faces and joining other burrows to form a great labyrinth. One colony contained 35 000 active nests along 12 km of cliff face! (Masello and Quillfeldt 2005).

Some of these birds – the pardalotes, as already mentioned, and some kingfishers, for instance – are only optional burrowers, using tree hollows at other times. Other birds also use other folks' burrows opportunistically. Australian Pratincoles hide chicks in rabbit burrows, presumably in place of those sadly no longer provided by bilbies or bettongs.

Dinosaurs in my garden

I don't think that any informed opinion doubts that the pardalotes and thornbills fluttering at the foliage outside my window are, for all meaningful purposes, dinosaurs. In 2007, a publication in the prestigious journal *Science* claimed to have extracted protein sequences from bones of a 68 million year old dinosaur (*Tyrannosaurus rex*), which are closest to birds among living animals (Asara *et al.* 2007). There was some disagreement about the reliability of this, but there is a mountain of other evidence, especially from the vast wealth of palaeontological treasures emerging in China, that birds arose among the running predators that gained infamy as the 'raptors' of Jurassic Park, and which were in the same dinosaur group as *Tyrannosaurus*.

Elsewhere, I have alluded to some of the essentially birdy traits that derived from their dino-ancestors: feathers (originally for insulation); the ability to maintain a constant body temperature (i.e. 'warm-bloodedness', or more properly homoeothermy); and the air sacs that enable a far more efficient lung function than our own.

Eggs themselves cannot be claimed in this category – they pre-dated even reptiles, let alone dinosaurs, when amphibians began to wrap their embryo in a parchment-like sac to enable them to break the shackles binding them to water. However, the quite different calcium-based shells that enclose bird eggs were certainly inherited from dinosaurs, as is the habit of brooding those eggs in a nest (initially on the ground) as opposed to burying them and leaving the hatchlings to make their way in the world. There are a surprising number of fossilised brooded clutches on which to base this claim (e.g. Norell *et al.* 1994, 1995). The realisation that eggs in exposed nests (i.e. brooded ones) are significantly less porous than those that are buried has greatly assisted in analysis of fossilised clutches (Tanaka *et al.* 2015). There is even evidence that Sensitive New Age Dinosaur dads took their turn at brooding (e.g. Birchard *et al.* 2013).

Chinese fossils have been found of dinosaurs that died in their sleep, with their heads tucked under their arms in the same posture as modern sleeping birds (Xu and Norell 2004).

Don't you love the concept that, even as we speak, a great great great (etc.) grand-niece of *Tyrannosaurus rex* is gleaning insects from a leaf near you?

Other memories of suburbia

Like moulted feathers drifting erratically down from the sky (will it land in my yard or not?), some unconnected thoughts of urban birds:

- In Guayaquil, the largest city in Ecuador, a crowded sprawling industrial town on the mighty Guayas River, a bright green and red cluster of Red-Masked Parakeets is noisily squabbling over a pipe extending from a building wall in the city centre, probably hoping to secure a nesting hollow. Sadly, this lovely boisterous bird is declining in its woodland habitats, partly because of loss of that habitat, but even more due to the voracious demands of the pet trade.

- Buenos Aires is a pulsing city of 3 million people. In the brief couple of days we were there, we didn't see a single stand-alone house, because everyone seems to live in high-rise apartment blocks. This is **dense** living! But, despite the constant swirl of people, we kept finding little parks, swamped in the roar of traffic, but supporting birds. A natural inhabitant of disturbed areas, the rusty-coloured Rufous Hornero has happily taken to the parks, to the extent that it has been declared Argentina's national bird! Attractive Rufous-bellied Thrushes join them on the grass, while above them speckle-necked pinkish Picazuro Pigeons have only recently discovered the benefits of urban life. In one especially seedy and busy little park, we even stumbled across a single Guira Cuckoo: they are known to adapt to urban living, but are renowned as highly sociable birds. A little mystery.
- Douala is another city in the 3 million club: the biggest city in Cameroon, and a serious shock to the system for someone unused to less-developed countries. My trip diary described it as 'vast, throbbing, noisy, frantic, hustling chaos', and the traffic as 'surreal lunacy'. Nonetheless, before we fled to the countryside we stopped in at a little area of gardens behind a razor-wired industrial complex by the huge Wouri River where a pair of African Grey Parrots flew overhead, flashing their red tails. Welcome to West Africa!
- Lima, the capital of Peru, is a huge throbbing intimidating mass of 10 million people, many of them poor rural residents who come to try to find work. One more placid area is that of Miraflores by the Pacific, which, because of the cold Humboldt Current onshore (and doubtless also pollution), is always shrouded in grey. Here, in a small park, I came across an Australian bottlebrush (*Melaleuca* sp., formerly *Callistemon*), flaunting big

red flower spikes. And availing itself of the nectar bounty was an Amazilia Hummingbird: a dryland green and rufous beauty with bright red bill, which has adapted to city living.

- Kampala, the Ugandan capital, is another rapidly growing crowded city, though its situation on the shores of Lake Victoria is ameliorating. The rooflines are strangely distorted by the huge louche shapes of menacing Marabou Storks resting between visits to the garbage dumps and abattoirs.
- An agitation of birds outside my study led me downstairs and cautiously into the backyard, where a little Collared Sparrowhawk glowered fiercely at me from the shadows of the tea-tree, daring me to take away the recently deceased House Sparrow clutched in her claws. Eventually she flew off with it to dine somewhere quieter (see Photo 26).

References

ABC (2017) Australian bird behaviour: have you spotted cockatoos raiding wheelie rubbish bins for food? ABC Radio Canberra, <http://www.abc.net.au/radio/canberra/programs/breakfast/ research-studying-rubbish-bin-opening-cockatoos-richard-major/8179948>.

Albright TP, Mutiibwa D, Gersond AR, Krabbe Smith E, Talbot WA, O'Neill JJ, *et al.* (2017) Mapping evaporative water loss in desert passerines reveals an expanding threat of lethal dehydration. *Proceedings of the National Academy of Sciences of the United States of America* **114**(9), 2283–2288. doi:10.1073/pnas.1613625114

Anon (2006) The urban evolution lab. *Biology News Net,* <http://www.biologynews.net/archives/2006/04/19/the_urban_evolution_lab.html>.

Asara JM, Schweitzer MH, Freimark LM, Phillips M, Cantley LC (2007) Protein sequences from Mastodon and *Tyrannosaurus rex* revealed by mass spectrometry. *Science* **316**(5822), 280–285.

Bancroft WJ, Roberts JD, Garkaklis MJ (2008) Vertebrate fauna associates of the Wedge-Tailed Shearwater, *Puffinus pacificus,* colonies of Rottnest Island: influence of an ecosystem engineer. *Papers and Proceedings of the Royal Society of Tasmania* **142**(1), 21–29.

Birchard GF, Ruta M, Deeming DC (2013) Evolution of parental incubation behaviour in dinosaurs cannot be inferred from clutch mass in birds. *Biology Letters* **9**(4), 20130036. doi:10.1098/rsbl.2013.0036

Blakers M, Davies SJJF, Reilly PN (1984) *The Atlas of Australian Birds.* Melbourne University Press, Melbourne.
Brooker MG, Brooker LC (1989) Cuckoo hosts in Australia. *Australian Zoological Review* **2**, 1–67.
Burton MW, Martin A (1976) Analysis of hybridisation between black-backed and white-backed magpies in south-eastern Australia. *Emu* **76**, 30–36. doi:10.1071/MU9760030
Chambers LE, Altwegg R, Barbraud C, Barnard P, Beaumont LJ, Crawford RJM, *et al.* (2013) Phenological changes in the southern hemisphere. *PLoS One* **8**(10), e75514. doi:10.1371/journal.pone.0075514
Dale S, Lifjeld JT, Rowe M (2015) Commonness and ecology, but not bigger brains, predict urban living in birds. *BMC Ecology* **15**, 12. doi:10.1186/s12898-015-0044-x
Dawkins R (2005) *The Ancestor's Tale: A Pilgrimage to the Dawn of Life.* Phoenix, London, UK.
Donovan T (2015) *Feral Cities: Adventures with Animals in the Urban Jungle.* Chicago Review Press, Chicago IL, USA.
Evans KL, Chamberlain DE, Hatchwell BJ, Gregory RD, Gaston KJ (2011) What makes an urban bird? *Global Change Biology* **17**(1), 32–44. doi:10.1111/j.1365-2486.2010.02247.x
Galeotti P, Rubolini D, Sacchi R, Fasola M (2009) Global changes and animal phenotypic responses: melanin-based plumage redness of scops owls increased with temperature and rainfall during the last century. *Biology Letters* **5**, 532–534. doi:10.1098/rsbl.2009.0207
Gardner JL, Heinsohn R, Joseph L (2009) Shifting latitudinal clines in avian body size correlate with global warming in Australian passerines. *Proceedings of the Royal Society B: Biological Sciences* **276**(1674), 3845–3852.
Gardner JL, Amano T, Sutherland WJ, Clayton M, Peters A (2016) Individual and demographic consequences of reduced body condition following repeated exposure to high temperatures. *Ecology* **97**(3), 786–795.
Garnett S, Crowley G (2017) 'National Recovery Plan for the Golden-shouldered Parrot (*Psephotus chrysopterygius*)'. Department of the Environment and Energy, Canberra, <https://www.environment.gov.au/node/15778#breed>.
Halfwerk W, Slabbekoorn H (2009) A behavioural mechanism explaining noise-dependent frequency use in urban birdsong. *Animal Behaviour* **78**(6), 1301–1307. doi:10.1016/j.anbehav.2009.09.015
Hollands D (1999) *Kingfishers and Kookaburras.* Reed New Holland, Sydney.
IPCC (2007) *Climatic Change 2007: Impacts, Adaptation and Vulnerability.* Contribution of Working Group II to the Fourth Assessment Report of the

Intergovernmental Panel on Climate Change. Cambridge University Press, Cambridge, UK.

Ives CD, Lentini PE, Threlfall CG, Ikin K, Shanahan DF, Garrard GE, *et al.* (2016) Cities are hotspots for threatened species. *Global Ecology and Biogeography* **25**, 117–126. doi:10.1111/geb.12404

Karell P, Ahola K, Karstinen T, Valkama J, Brommer JE (2011) Climate change drives microevolution in a wild bird. *Nature Communications* **2**(1), 208. doi:10.1038/ncomms1213

Kaufman L (2011) Conspiracies don't kill birds. People, however, do. *New York Times*, 17 January 2011, <http://www.nytimes.com/2011/01/18/science/18birds.html>.

Maklakov AA, Immler S, Gonzalez-Voyer A, Rönn J, Kolm N (2011) Brains and the city: big-brained passerine birds succeed in urban environments. *Biology Letters* 7(5), 730–732. doi:10.1098/rsbl.2011.0341

Masello JF, Quillfeldt P (2005) A conservation update on the burrowing parrots of Argentina. *Parrots* **92**, 36–40.

Mitterdorfer B (1996) A morphological and genetic study of the Australian Magpie (*Gymnorhina tibicen*) hybrid zone. BSc (Hons) thesis, Australian National University, Canberra.

Norell MA, Clark JM, Demberelyin D, Rhinchen B, Chiappe LM, Davidson AR, *et al.* (1994) A theropod dinosaur embryo and the affinities of the Flaming Cliffs dinosaur eggs. *Science* **266**(5186), 779–782. doi:10.1126/science.266.5186.779

Norell MA, Clark JM, Chiappe LM, Dashzeveg D (1995) A nesting dinosaur. *Nature* **378**, 774–776. doi:10.1038/378774a0

Parmesan C, Yohe G (2003) A globally coherent fingerprint of climate change impacts across natural systems. *Nature* **421**, 37–42. doi:10.1038/nature01286

Pizzey D, Doyle R (1980) *A Field Guide to the Birds of Australia.* Collins, Sydney.

Potvin DA, Parris KM, Mulder AM (2011) Geographically pervasive effects of urban noise on frequency and syllable rate of songs and calls in Silvereyes (*Zosterops lateralis*). *Proceedings of the Royal Society B: Biological Sciences* **278**(1717), 2464–2469.

Reinberger S (2013) Birds that go wild for the city. *MaxPlankResearch* **1**(13), 72–79.

Russell A (1953) *Murray Walkabout.* Melbourne University Press, Melbourne.

Tanaka K, Zelenitsky DK, Therrien F (2015) Eggshell porosity provides insight on evolution of nesting in dinosaurs. *PLoS One* **10**(11), e0142829. doi:10.1371/journal.pone.0142829

Woinarski J, Bonan A (2017) Pardalotes (*Pardalotidae*). In *Handbook of the Birds of the World Alive.* (Eds J del Hoyo, A Elliott, J Sargatal, DA Christie and E de Juana). Lynx Edicions, Barcelona, Spain, <http://www.hbw.com/node/52350>.

Woodall PF (2017) Kingfishers (*Alcedinidae*). In *Handbook of the Birds of the World Alive.* (Eds J del Hoyo, A Elliott, J Sargatal, DA Christie and E de Juana). Lynx Edicions, Barcelona, Spain, <http://www.hbw.com/node/52271>.

World Wide Words (2001) Stool pigeon. World Wide Words website, <http://www.worldwidewords.org/qa/qa-sto2.htm>.

Xu X, Norell MA (2004) A new troodontid dinosaur from China with avian-like sleeping posture. *Nature* **431**, 838–841. doi:10.1038/nature02898

7
Woodlands and grasslands

Great Western Woodlands, Western Australia: Emu family

They materialised by the road east of Hyden in the southern inland of Western Australia when we were almost alongside them. The big male Emu watched us carefully as we pulled over, while the youngsters seemed largely heedless of any potential hazard – I guess they were used to taking their cues from their father. Stretching up to appear more intimidating, he stood most of 2 m high, his sooty brown body somewhat reminiscent of a haystack of loose shaggy feathers, and the short black fluff of head and neck framed a large patch of blue skin around his ears and down his neck. The dozen largish chicks, a couple of months old, were marked with creamy stripes running down the neck and along the body, like a flock of big bumblebees. They hovered at the edge of the Gimlet Gums that reared with oddly fluted glowing coppery trunks from the understorey of scattered shrubs and grasses.

These Great Western Woodlands of the semi-arid Goldfields district of Western Australia represent the largest extent of Mediterranean climate woodland in the world that remain in their original state (Mediterranean implying a hot dry summer and winter rainfall). They cover 16 million hectares: an area greater than that of England and Wales combined. The apparently – and understandably – mandatory drive for Australians, 4000 km 'across the Nullarbor' from the east coast to Perth on the south-west coast, passes through these woodlands, but a quieter 300 km journey along the well-maintained Granite and Woodlands Discovery Track between Hyden and Norseman is a more rewarding way to see them (see Photo 27).

The term 'woodland' is used loosely in day to day conversation and, even more confusingly, is employed differently in various parts of the world, but the essence of the concept in most places is a treed habitat where the trees are more sparsely scattered than in a forest; that is how I apply it in this chapter. In Australia the definition generally specifies that the total tree canopy from above covers between 10% and 30% of the ground area, meaning that the individual canopies scarcely touch, if at all; if more than that it is a forest. In Britain, however, 'woodland' is broadly used to describe any natural habitat dominated by trees, while 'forest' implies a managed plantation. There are other important implications of this scattered nature of the woodland tree cover too, the chief one being that much more sunlight hits the ground than in a shady forest, so that the understorey is quite different. Most grasses like to grow in full sun and a woodland understorey comprises more grass and fewer shrubs than a forest (bearing in mind that all these habitat concepts are better thought of as a continuum rather than discrete separate entities). As conditions change, so that tree growth becomes less and less supportable (perhaps because of decreasing rainfall or temperature, or changes in soil type, such as to cracking black soil plains whose constant expansion and contraction rips roots apart) we are left with just the woodland understorey: a grassland with few or no trees.

I tried waving a handkerchief out the side window, which can often attract curious Emus to come close to investigate (a curiosity which in times past was exploited in various ways by Indigenous hunters), but, as they started to move cautiously towards us, another vehicle roared past in a cloud of dust, seemingly oblivious to the birds, and the Emus fled. He jumped and whirled in the air, the haystack morphing into athletic dancer, pounding down the road verge seemingly powered by little puffs of dust with each stride. The chicks followed as he eventually swerved off into the trees: it often seems to take Emus a while to work out that they can actually leave the road.

Ratites, and the mysterious case of the flying elephant birds

The ratites, as traditionally defined, are quintessentially ancient Gondwanans, including all the great flightless birds of the southern continents – two species of ostrich in Africa, the Emu and three cassowaries in Australia–New Guinea and three (or perhaps just two) rheas in South America. They also include the five New Zealand kiwi species, which, of course, are much smaller. In addition, there are several recently extinct species (i.e. in the last few centuries, coinciding with human arrivals) comprising perhaps four, and possibly up to seven or eight, species of mighty Madagascan elephant birds, two emus (though here, as with the elephant birds, taxonomy is uncertain) and nine New Zealand moas (see Photo 28).

The most fundamental division of living birds is not, as one might suppose, into passerines and non-passerines (see page 47), but into the Palaeognaths and Neognaths. These reflect palate characteristics, and the division has long been recognised from anatomical studies: an analysis more recently confirmed by biochemical and genetic studies. The 'ancient palates' have generally been understood to comprise the ratites, plus an associated 'sister group', the South American tinamous (see page 42): 47 species of smaller ground-dwelling birds that fly, albeit weakly. The conventional wisdom has been that the ratites lost their flight back in the hazy mirages of Gondwanan time, and that the flightless ancestors of modern ratites were carried across the Southern Hemisphere with the fracturing of Gondwana into rafting continents. Along with the flightless rhea ancestors to South America went the cousins, the still-flying tinamous. Knowledge of the history of the component continents of Gondwana (in particular, the order in which they broke away from each other) and logic tell us that ostriches and elephant birds should be the oldest members of the group and be most closely related to each other. Likewise, moas and kiwis, stranded together as New Zealand floated away, must surely be each other's nearest relatives. Moreover, although the tinamous are undoubtedly related to the ratites, they can't actually *be* ratites because that would imply that they had somehow regained flight after being isolated in South America, and it is universally agreed that such a reversal of evolution is simply impossible (too many separate adaptations to a radically new lifestyle for all to be wound back in synch, for a start).

But … again, that 'but' to send shivers down the spine of those of us fond of a neat story. The story I've just summarised, and which I've

Which DNA to analyse?

Cell mitochondria contain much less DNA than does the cell nucleus, so it is relatively easy (but still not that easy!) to obtain a complete mitochondrial analysis. Moreover, animal mitochondrial DNA evolves more rapidly than does that of the nucleus. These two observations mean that doing complete mitochondrial analyses to compare species is now a standard and powerful technique for unravelling not only relationships, but the time since their 'most recent common ancestor' stalked the Earth: indeed the acronym MRCA is now in widespread use in the literature without perceived need of clarification. Such analyses have rapidly become almost standard practice and have been applied to a huge array of organisms – just try entering 'complete mitochondrial analysis' into your search engine of choice. Moreover, as one of the key papers of this ratite evolution revolution notes, 'ancient DNA is now a respectable and thriving industry' (Phillips *et al.* 2010). DNA can be retrieved from fossil material (as indeed any aficionado of Jurassic Park already knew!) so that extinct pieces of the puzzle, such as elephant birds and moas, can now be fitted into place. And as it has turned out, they complete a most unexpected picture.

told in good faith many times over the years, *seemed* to explain all that we observe about the ratites, and to encapsulate the role of Gondwana in explaining what we see in the Southern Hemisphere. Except that now it seems not to do so after all ... There have for some time been mutterings of unease about the narrative, not least because, although it is agreed that the various far-flung outliers of the palaeognaths are indeed old, surely they can't be *that* old? And sure enough, in the past decade a series of studies using emerging techniques have completely disassembled the story and rebuilt it from scratch, based on the apparently incontrovertible new information that seems to accumulate by the year.

For instance, it emerges that the tinamous, rather than being convenient 'outliers' to the main ratite line, are right in the middle of it – oops. The 'oops' is because this means that the common ancestor they all share was either flightless, which is the conventional wisdom, and the tinamous, against all conceivable probabilities, regained flight (and no-one believes that), or it flew and the others all then lost their flight independently!

It gets worse, though, from the viewpoint of those wishing to cling to that conventional wisdom: the closest relatives of tinamous are not rheas at all, but … New Zealand moas! (To be honest, others had suggested this previously, without being able to explain the distributions.) Moreover, kiwis and Madagascan elephant birds are each other's nearest and dearest; this was new. (I've given up on the exclamation marks now – it all seems a bit *Alice in Wonderland* really.) A series of papers has developed this theme, naturally with a lot more detail and sophistication than I am reflecting (e.g. Harshman *et al.* 2008; Phillips *et al.* 2010; Allentoft and Rawlence 2012; Baker *et al.* 2014; Mitchell *et al.* 2014).

I confess that, when I first came across this, my reaction was that there must have been some mistake. I'm a firm believer in the parsimony principle: that the simplest evidence-based explanation is always likely to be the correct one. The more evolutionary steps that are required to explain something, the less likely it is that it happened that way. You see, the initially somewhat hallucinogenic implication of all this is that at least some of the ancestral rheas, tinamous, kiwis and elephant birds flew in to their current abodes, and independently subsequently all (except the kiwis and tinamous) grew hugely in stature – they couldn't have flown at their current size – and lost their ability to fly. (Or it could have been the other way round: that is, they could have lost their powers of flight and then grew.)

In fact, when the analysis was done of the time since the lines separated, and that was compared with the timing of the breakup of Gondwana, it emerges that *all* of them must have flown in, because each pair separated well after their current isolated continent was last connected to others. For instance, kiwis and emus/cassowaries only parted ways some 60 million years ago, but New Zealand had become isolated 20 million years before that. (Phillips and colleagues, who did this study, didn't have the reliable elephant bird material that was available to Mitchell's team just 4 years later, so couldn't make the elephant bird–kiwi connection.) Moas and tinamous separated at about the same time (i.e. 60 million years ago). Emus and cassowaries diverged only ~20 million years ago, but were already on the same continent.

Finally, in terms of shocks, it emerges that ratites might not even be Gondwanan in origin, though that is still uncertain. Ostriches seem to have arisen as a separate line in the earliest ratite days, but whether in southern Gondwana (where the ancestor of all the rest apparently dwelled), or actually in Eurasia where they later lived, is unclear. Certainly, it appears from the fossil record that ostriches may only have turned up in Africa as recently as 23 million years ago.

So how do the authors go about justifying the apparently impossible coincidence of all these birds independently abandoning their wings at about the same time and getting very big indeed? What could possibly have happened across the Southern Hemisphere to explain such a seemingly uber-implausible set of events? Well, something pretty cataclysmic actually did happen about then – a massive meteorite smashed into the Earth on what is now Mexico's Yucatán Peninsula 65 million years ago and, in the ensuing deep chill as the sun was blocked out for years by the dense blanket of dust and smoke, three-quarters of all animal and plant species on Earth perished. It was a horrific time as life chilled and starved, and was burnt by widespread acid rains. Among those extinguished were all the dinosaurs, that mighty dynasty that had dominated the planet for 160 million years; well, not quite all, of course, because the bird-dinosaurs survived and thrived in the suddenly empty landscape.

Among the niches left vacant was that formerly occupied by the birds' immediate ancestors: the fast, erect bipedal dromaeosaurs. We know that modern birds are very prone to evolving to flightlessness in situations where there is no reason to go on spending so extravagantly to maintain an aerial lifestyle (especially on islands; see page 69). Perhaps then it is not so implausible, in such a different empty world where opportunities were many and predators were few, that a series of related birds should do so in isolation from each other, having increasingly relied on their legs, rather than their wings, which were far more expensive to fuel. Presumably too they still carried the genes of their bipedal running ancestors. Freed of the weight restriction imposed by flight, they could then grow increasingly bigger to tower over such predators as were also evolving. Phillips *et al.* (2010) point to the analogy of Australasian Swamphens flying (or being blown) from Australia to New Zealand in relatively recent times, and in that mammal-free Nirvana losing the use of their wings to evolve into the big flightless Takahe.

Maybe it's time to come back to now and the Emus, which really don't care much about ancestry.com.

Feathers: a bird's best friends

I mentioned the curiously 'haystackish' appearance of the Emus, whose plumage is more reminiscent of an untrimmed shaggy dog.

Feathers are critical to a bird: they help determine where it lives and how it lives. A measure of their significance is that, despite their proverbial lightness, and the fundamentally critical importance of shedding every gram of surplus weight in the interests of flight, a flying bird's feathers weigh two to three times as much as its skeleton.

A feather is formed of keratin, which is unsurprising given that it is the protein that forms reptile scales, from which feathers evolved. There are two major feather types: vaned feathers and down feathers. Vaned feathers are essentially all the visible ones, flight and tail feathers, and all the body-covering contour feathers. They consist of a solid shaft, or rachis, arising from a hollow shaft, or calamus, which is embedded in the skin. From the rachis extends the vane (i.e. what we would probably think of as the feather), comprising two densely packed opposite rows of barbs. In most visible feathers, these barbs are locked to the ones alongside by lines of tiny hooked barbules, rather like velcro. Birds spend hours a day preening, meticulously running every feather through their beak to reset the barbules, making sure they're properly zipped up so they can perform their critical functions of flight, waterproofing and insulation.

At the base of many feathers is a short non-zippable woolly aftershaft (or hyporachis, if you're trying to impress someone), like another little feather growing from the base. Its function is presumably insulation, though not all birds have them. Based on the bird groups that do and don't have them, it seems to be a primitive characteristic, with passerines (the most recent Order of birds to arise) mostly lacking aftershafts. At the other extreme, in our ancient roadside Emus and the closely related rainforest cassowaries, the aftershaft is as long as the main shaft, so each feather appears duplicated. Other ratites don't share the extended aftershaft, but all lack barbules so the feathers don't lock together, hence the loose shaggy or even woolly appearance.

Other modern birds have an array of other feather types too, some of them very specialised. The other common feather type, as mentioned above, is the down feather: the soft fluffy doona that lies beneath the vaned feathers and keeps the bird warm. Chicks hatch with only down, with the vaned feathers growing through later. Down feathers have a very short rachis, or none at all, and no barbules to zip the fluffy barbs together. Of course, it is not strictly the feather itself that insulates but the layer of air it traps – like double glazing. A bird can increase insulation by means of tiny muscles just under the skin that can raise feathers – 'fluffing up the doona' – or can decrease it in hot weather by flattening the feathers against the body to squeeze out the air and increase conduction of heat away. We have all seen birds on a cold day looking twice their normal size, with feathers puffed out all over them.

Bristles around the eyes of some species, especially insect-eaters, consist of just a thin shaft. They are thought to be sensory, but probably also play a role in protecting the eyes from struggling prey that are understandably less than enthusiastic about being eaten. It seems that they play no role in actual prey capture, though this has also been proposed (Lederer 1972). Filoplumes are also single shafts, but with lots of sensors at the tip. They are scattered through the plumage, but especially among the flight feathers, and apparently convey information to the brain about feather movement so that adjustments can be made. These remarkable little structures can be seen as the 'hairs' on the body of a plucked chook.

Powder down is another pretty amazing feather type, present in a minority of bird Families, though these are as diverse as pigeons, parrots, cotingas, herons, frogmouths and the strange New Caledonian endemic Kagu. In parrots and pigeons, the powder down feathers grow in little thickets on different parts of the body, while in some other groups they are scattered throughout the plumage. These different patterns apparently reflect the fact that they evolved independently in different families. The feathers have tips that disintegrate into keratin powder, like talc (though not chemically), which either disperses through the feathers naturally or is spread with the bill tip. Either way, it apparently cleans, conditions and waterproofs the feathers (can't you imagine the TV ad for its remarkable properties?). Another interesting characteristic of powder down feathers is that they are not moulted, but grow continually to replace the eroding tip (see Photo 29).

Despite the lavish care given them, feathers inevitably deteriorate. They get worn by collisions with dust particles in flight, by the normal wear and tear of a vigorous life, by ultraviolet deterioration, by some bacteria that have evolved to digest feather protein and by the insidious feather lice. These are remarkable little animals that are among the ultimate specialists: not only is a feather louse likely to be limited to just one bird species, but to one section of its body.

Moreover, a feather, like a hair or a fingernail, is dead, so once damaged it can't repair itself (though hair or nail will continue to grow; a mature feather does not). The solution is moulting: the complete sequential replacement of all the feathers at least once a year.

Moulting is a very tricky and potentially hazardous process: you can't have gaps that leave you without proper body insulation or adequate flying abilities. Generally the process on one side of the body is mirrored on the other: for example, one feather on each wing, a small patch on each flank. The old feather is pushed out by the new, which grows from the same follicle. The process normally starts in autumn, when the most recent brood has been waved goodbye from the territory, or at least are finally looking after themselves. That very stressful period of feeding growing and demanding youngsters doesn't need to be made more so by dealing with the handicap of incomplete plumage at the same time. It can take weeks, or even months, to complete the process, though some birds that have strategies to cope with a flightless period can speed it up. Many waterbirds, including ducks and swans and some seabirds, go to a nice safe quiet refuge, such as a lake or an island, shed all their flight feathers at once and are back to full flight capacity in 3 or 4 weeks time.

Immature birds must moult into their adult finery, while those species in which the males (and, in a few instances, the females) stay incognito for much of the year but adopt colourful breeding attire, such as shorebirds, weavers and fairy-wrens, must moult their body feathers (though not those of wings and tails) twice a year. The Splendid Fairy-wren (now there's an unhelpful name, among the galaxy of other splendid fairy-wrens!) that we'd admired further back along the road east of Hyden was just coming into his courtship glory. His head and neck were already flaunting a mix of patches of brilliant sky blue and deepest navy, but the rest of him was still drab winter brown with a scattering of blue feathers: presumably having a blue head is more important than the rest of him.

And always, it seems, there are trade-offs to be made. If a bird finishes breeding late in the season – perhaps because it delayed starting until conditions were more conducive, or took a chance on a second clutch – it must hurry its moult to be at peak flying capacity and optimal insulation before winter strikes. However, it transpires that hurried feathers are not very good feathers. House Sparrows with accelerated moult, prompted by messing artificially with their day length, had shorter flight feathers with more 'fault bars' (weaknesses that can lead to breakages in barbules and even the main shaft) and body feathers with reduced insulating capacity (Vágási *et al.* 2012). If the birds were in poor condition to start with – perhaps after a difficult breeding season – the feathers were even less effective. The moral seems to be that if you're not going to moult early, you'd better be in good condition; if not, you're likely to find yourself in a nasty spiral that continues over coming seasons.

But there was another attention-catching aspect of the running Emus too ...

A bird's leg: a wonder of redesign

An obvious thing when watching the elegant power of a running Emu is to make comparisons with human sprinting: not so much the elegance and power as the very basis of it – standing erect and using just two limbs for propulsion. But – yes, yet again that 'but'...

Something is very different: an Emu's knee bends the 'wrong' way (well, wrong for a human leg at least). What on earth would have been the point of, at some stage of either our or their development, turning the leg completely back to front from its reptilian origins? Furthermore, it would have been the birds' legs that were reversed, because a modern lizard's leg still bends the same way that ours does.

The answer is that it's a trick question: a bird's knee bends the same way that ours does, *but we are not looking at the actual knee when we make the judgement.* Over and over again, the answer to a puzzle about some aspect of a bird's structure or behaviour comes back to necessary compromises to maximise flying efficiency, which in turn mostly comes down to weight minimisation. Not in this case, however – though not because the Emu gave up flying long ago, because those basics of anatomy go way back to its flying ancestors and are shared with all other birds (think of any other bird you like). No, the flight-related compromise in this case is driven by the great flight muscles that take up so much of the front part of the body. There is no room for the hips to be underneath the bird's centre of gravity, and a bird doesn't normally stand bolt upright but leans forwards, perhaps for reasons of vision and foraging and because of the attachment angles of the wings. The attachment point of the leg to the spine, the hip, is right at the back of the body from where gravity should cause the bird to fall flat on its beak.

However, something very clever has happened in the past to compensate for this. The thigh, instead of projecting down from the body as ours does, is relatively short and runs forwards along the bird's flanks; it is entirely covered by skin and feathers and anchored immovably in place by muscles against the side of the bird. We don't see it at all in a living bird, but you only need to think of a roast chook to realise what I mean. When carving, the thigh must be cut out of the body.

Now the thigh ends pretty much under the centre of gravity, so the leg may safely leave the body – but this joint (functionally a 'hip') is in fact the true knee, being at the end of the thigh bone. *And this knee really does bend backwards*, just as ours does! So the *apparent* 'thigh' of the bird leaving the body is in fact the true shin, and the apparently forward-directing 'knee' at the lower end of it is really the ankle ... To make it even trickier (only for us, of course, this is not a problem for the bird) – the knee/'hip' and the true shin/'thigh' below it are often covered by feathers in all but the biggest birds, and are also often hidden by the wing.

I hope you haven't given up in exasperation by now, because I reckon this is really interesting. But you might instead have given up because I'm obviously talking nonsense – after all, below this 'knee' that I'm calling an ankle there is another section of leg that looks and functions for all the world as a shin. Fair enough too. However, this is another very clever evolutionary solution and we couldn't be expected to recognise a piece of leg that we don't need and for which we thus have no equivalent. It is comprised of a mix of extended ankle bones from above and foot bones from below, and it does just fine the job of keeping the bird at a sensible distance from the ground. Because most of the foot bones have been co-opted to make this new 'shin', birds have no choice but to (almost exclusively) walk – and run – on their toes.

Meanwhile, the Emus have certainly had it away on their toes, so time for us to move on too.

Extinction

The Regent Honeyeater is a truly, but oddly, lovely woodland bird. John Gould thought it 'one of the most beautiful birds inhabiting Australia' (Gould 1865). It has a black head and neck, with a strange warty bare patch around the eye, and the rest of the body is chequered and scaled in a complexity of black and yellow and white. It is also terrifyingly rare, seemingly fading out of existence before our eyes, even as lots of good people are working with increasing desperation to stop that happening – rather like watching a medical team fighting a steadily losing battle in the emergency department.

I first saw a Regent in 1992: a pair of them busily gathering spider webs from the black deeply incised trunk of a woodland ironbark eucalypt at Wyangala Dam north of Canberra. Unfortunately, the group I was accompanying had very little interest and wandered off after a few seconds, but I was enthralled. I've seen them just three times since then, all in Canberra: a pair nesting in woodland at Mount Majura in 1994; then a decade-long hiatus until a single bird appeared not far from there on Mount Ainslie; and another single on the Australian

National University campus in 2005. Not one since then. Unfortunately, the temperate grassy woodlands of south-eastern Australia in particular are among our most damaged and least protected habitats.

Perhaps the worst part is that the Regent Honeyeater used to be not only common, but abundant, as revealed by accounts from around the turn of the 20th century: for example, it 'moves about the country in flocks from 50 to 100 or more individuals' (Australian Town and Country Journal 1896). In September of 1909, a report from Belltrees near Scone in the Hunter Valley claimed that, 'ever since March last they have been with us in thousands' (White 1909). And now? Probably less than 400 mature individuals in the whole world, and still decreasing, scattered ever so sparsely from eastern Victoria to northern New South Wales, though they once spread west to Adelaide and north into Queensland (Garnett *et al.* 2011; ANU 2016).

I passionately want my pessimism over the Regent to be wrong: I shall genuinely grieve if it becomes the first mainland Australian bird species to become extinct in the wild in nearly 100 years. The last one was the Paradise Parrot of south-eastern Queensland in the 1920s and, if the Regent follows it, we can be reasonably sure it won't be the last, especially among the woodland birds. The White Box-Yellow Box-Red Gum woodlands have been comprehensively destroyed for agriculture and urban development; one estimate is that less than 5% remain (ACT Government 2004). However, attempts to assess the remnants which are in anything like an original state are even more alarming – one well regarded study on the White Box woodland component concluded that only 0.1% of the original habitat survives substantially intact (Thiele and Prober 2000).

But why the Regent in particular? Other woodland species are struggling, but probably none are in such a catastrophic situation. Sadly, we can't really say definitively, though we understand various factors that, in combination, doubtless contribute. They seem to need particularly high-quality habitat, and wander the countryside apparently seeking areas of continuous high nectar flow. Sadly, the remaining woodland remnants mostly comprise lower quality patches unattractive for agriculture, and represent marginal Regent habitat.

Some other large honeyeaters are more generalist in their requirements, and move into disturbed woodland. The closely related Red Wattlebird is larger and highly aggressive, and the slightly smaller Noisy and Yellow-throated Miners live in tough bullying xenophobic gangs. Gould made special mention of the Regents' pugnacious disposition, but now they are overwhelmingly outnumbered wherever the living is good. And how do you help a species comprising just a few hundred birds that are constantly on the move over tens of thousands of square kilometres of countryside, and that can be absent for years at a time from a previously favoured locale? We don't really know, other than via the current superhuman efforts at revegetating and monitoring. We still don't know for sure why we managed to lose the Paradise Parrot and not others.

It is perhaps fortunate that the approach to assisting more sedentary species is (at least theoretically) more straightforward. For woodland species, for instance, we need to understand the minimum requirements of the threatened species in terms of: the minimum size of a remnant vegetation patch; the maximum distance to the next patch; key tree, shrub and herb species; the nature of the understorey; the age of the trees; and so on. Often it is possible then to identify an 'indicator' or 'umbrella' species whose requirements are such that providing them will automatically protect other species. A useful definition is 'a species whose presence indicates the presence of a set of other species and whose absence indicates the lack of that entire set of species' (Lindenmayer *et al.* 2000). Sometimes a large predator (e.g. a Jaguar or large forest eagle in South American rainforests) is cast in that role, though this is not likely in an Australian woodland. However, in the south-eastern Australian woodlands, both Brown Treecreepers and Hooded Robins are reliable indicator species – if they are present, all other threatened woodland birds are also likely to be. Unfortunately, Regent Honeyeaters don't fit this model.

Extinction, like evolution, is not an event (other than for vanishingly rare broad-scale accidents such as meteorite strikes) but a process. I was once taken publicly to task by a journalist who objected to my use of 'extinction' to describe the loss of Brolgas from the Canberra region –

'radically devaluing' the word was the accusation. He insisted that it could only be used for the final and total loss of a species and that its use for anything less would alarm people to the point that they wouldn't be appropriately alarmed when something worse came along. He climaxed by accusing me of 'humpty-dumptying'. It was a nice insult, but I believed it was misapplied. We may choose to define extinction as the moment that the last individual disappears – and for a few species where the last individual was known to be in a zoo (the Passenger Pigeon and Thylacine spring accusingly to mind) we can indeed precisely log the dismal moment. If, however, we reserve use of the word only to report that the organism has gone forever, that seems to me to be a hollow, and perhaps even amoral, usage. We really do need to be alarmed, and to respond accordingly, well before that happens.

I think of all the populations that comprise a species as being like little lights across the countryside. Extinction begins when those little lights start to flicker and go out until finally – total darkness. We should see the spreading tenebrous patches as our alarm call, rather than regard the remaining lights as reason for complacency.

There are other aspects too. Although ultimately the survival of the species is paramount, its disappearance from areas where it used to live is cause for genuine sorrow, as well as concern for the broader implications. The plains around Canberra supported flocks of Emus, Brolgas and Australian Bustards. The woodlands rang at night with the thrilling wailing choruses of Bush Stone-curlews: I have a friend who remembers them calling in northern Canberra in the 1950s. No more though. The last three species named can no longer be found in any numbers within many hundreds of kilometres of Canberra, though they are still quite common in the tropics. Does that make it acceptable though?

Another serious problem is that we are still unravelling bird taxonomy all over the world. In Australia, over the last decade or so, around 16 new species have been recognised (nearly all passerines), mostly unanimously among taxonomists, through examination of separate populations previously assumed to have belonged to one species. These include three new species of quail-thrush (nearly doubling the number), and Western Australian populations of common eastern states species the White-naped Honeyeater and Golden Whistler (e.g. Dolman and Joseph 2015). What if some of those populations had been allowed to slip into oblivion under the assumption that the 'species' was doing well enough elsewhere?

Often the cause of extinction is clear enough. On the Limestone Plains (where Canberra now stands), the Emus and Australian Bustards were hunted to extinction. The Brolgas were shot too, but drainage of wetlands was probably a more profound issue. The stone-curlews were eaten by foxes, in addition to the loss of woodlands. Rarely, however, has the cause been so explicitly tied to a single incident as the shipwreck of the *Makambo* off Lord Howe Island, which introduced rats to the island (page 66), or to the moment when the Duke of York stuck Huia feathers into his hatband.

The Huia was a remarkable bird; indeed all birds are, but this one really was special, and of great spiritual significance to the Maori. It belonged to a small New Zealand endemic Family called the New Zealand wattlebirds (no relation to the Australian honeyeaters of the same name). It was a big darkly iridescent bird of the deep wet forests, with broad white tail tip and large red face wattles. But the bill! Or rather bills, because those of male and female were quite different. His was heavy and crow-like, as the descriptions put it; hers was long, slender and deeply decurved. It seems they worked together, he breaking open the surface of rotting logs, and she probing deeply into the loose material to extract big juicy grubs. Well, that's one story anyway, though, like so much else about the Huia, it's hard to be certain. Perhaps they simply focussed on different foods, including fruits, to enable them to live in a smaller territory.

Living in old growth forests on the North Island, they suffered from habitat loss from the time the Maori arrived, but things got a lot worse in the 19th century as vast swathes of forests were turned to farmland. The Maori hunted Huias for feathers, but only very senior people could wear them and there were firm restrictions on when they could be taken. European settlers, however, slaughtered them to sell to private collectors and museums back in Europe. Nonetheless, by the end of the 19th century, the bird survived in mountain forests and it is possible that the new conservation laws would have in time been better enforced than they were (essentially they were blatantly ignored).

Then, in 1901, the Duke of York (later to become Britain's King George V) visited New Zealand and at Rotorua the local Maori presented him with a Huia feather as a sign of respect: he put it in his hatband and at that moment the fate of the spectacular bird was sealed. 'Huia plumes were reduced from sacred treasure to fashion accessory' (Szabo 1993). The price of an individual feather was £1, so a single bird was worth £12 – at a time when a labourer could expect to earn between £2–3 a week. Organised shooting parties went out and shot every Huia they could find. Around road and rail construction camps hundreds of birds were killed. As the birds got ever scarcer, prices for a single feather rose to £5. Moreover, rather than spur the desperate rescue efforts that we might hope to see today, the attitude seemed increasingly to be one of 'it's doomed anyway, so we might as well make what we can out of it'. The last confirmed sighting of the bird was in 1907, just 6 years after the duke's fatal gesture. This was effectively a planned extermination, abetted by government: islands were set aside as reserves, but no birds reached them. Indeed, a pair destined to be the first were diverted as a gift (dead of course) to Lord Rothschild in England (Szabo 1993; Fuller 2002; Higgins *et al.* 2006).

Most extinctions don't happen like that, though in past years a lack of concern was usually a hallmark. The cause (or causes, there's rarely just one) was clear, but that isn't always the case either. I alluded earlier to two emu species that have become extinct since European settlement. One was from Kangaroo Island, off the coast of South Australia, the other was from King Island in Bass Strait between the mainland and Tasmania. Both island emu forms were regarded as 'dwarf', with the Bass Strait birds being smaller (about half the size of modern mainland Emus). Both were reported to be very common, but barely survived the initial contact with Europeans – in both cases represented initially by whalers and sealers keen to supplement their meat supply. The great French scientific expedition led by Nicolas Baudin visited both islands in 1802, and captured live birds from each. Unfortunately, the science was a bit rubbery on those occasions and proper records were not kept of which came from where. He took on board an unspecified number (apparently four or five) at King Island, and two from Kangaroo Island, but they were allowed to mingle; only two reached France, where they lived for another 19 years to be perhaps the last survivors of their kind (West-Sooby 2013).

In a sense it is immaterial to the story, but the confusion thus engendered is worth mentioning. It was originally assumed that all the birds that arrived alive in France came from Kangaroo Island. When someone finally got round to studying the skeletons of the two birds, it transpired that they were different from each other, and the specimen on which the Kangaroo Island Emu description was based (a skeleton in the Paris Musée d'Histoire Naturelle), was most likely from King Island, so things got tricky. It fell to Shane Parker of the South Australian Museum to write a new description of the species (as he asserted that it was) based on subfossil leg bones from the island: he called it *Dromaius baudinianus* to honour Nicolas Baudin (Parker 1984). Today there is far from unanimity on the species status of the King and Kangaroo Island Emus: the Australian Government regards them as separate species (Department of the Environment 2017a, 2017b), while a mitochondrial DNA study of King Island Emu subfossil remains concluded that they fell well within the variation expected from mainland birds (Heupink *et al.* 2011). Perhaps it doesn't matter nearly as much as the fact that these magnificent animals – the products of at least tens of thousands of years of adapting and evolving in isolation – were crushed from existence and wiped from the record virtually overnight, and we can't even say why.

Astonishingly the King Island Emus were gone from the planet by 1805, just 3 years after Baudin reported them in abundance. Certainly the whalers were hunting them with dogs, but King Island is 60 km long and over 20 km wide and it seems unlikely that this alone could have resulted in extinction in such a short time. Even more surprising is the total loss of the Kangaroo Island Emus: it took a little longer, but they had gone by 1827. My surprise here is based on the size of Kangaroo Island, which is 145 km long and 45 km wide, and heavily vegetated. It simply seems too hard to accept that hunting alone could have achieved their complete destruction in that time, especially bearing in mind that kangaroos were also being slaughtered there by the thousands and are still common to this day. Perhaps the most likely explanation lies in a greatly increased burning regime, which could well have been fatal to a ground-nesting bird. We will simply never know: all we can do is wonder and mourn.

Perhaps it's time to think of things less grim (though we can't, and shouldn't, ever fully forget what has been and is being done to the world by our cupidity, arrogance and ignorance – always a toxic brew).

High Veld, Wakkerstroom, South Africa: Widowbirds

It was a remarkable performance, and one that required some suspension of disbelief to absorb. A black streamer fluttered slowly above the grasslands: it took a bit of effort to see it as a bird. My diary at the time described them as resembling 'overdecorated runaway toy kites'. The Long-tailed Widowbird (yes, a male widow) flapped slowly, showing off his glowing red and white shoulder 'epaulets', but the real show-stopper was the tail: the streamer that first caught our attention. He is not much bigger than a House Sparrow, with a wing span of some 15 cm – but the tail is close to half a metre long! The long feathers hang loosely, forming vertically flattened pennants dangling down behind him.

Wakkerstroom, a busy little village some 230 km south-east of Johannesburg, lies in the vast flat expanse of the high veld: that great grassy plateau in east-central South Africa. It is nearly 1800 m above sea level, which means winter can be tough (freezing at night and still cold during the day; frost and snow are common). Wakkerstroom started out as a 19th-century service centre for the local agricultural communities. (Actually it started out as Uysenburg and went through a couple of name changes before settling on Wakkerstroom in 1904; it was an attempt to acknowledge the Zulu name of the local river – but translated into Afrikaans!) Today it makes a great deal of its living by catering to the birdwatchers who flock to it, appropriately, to enjoy the rich and special birdlife of the grasslands and wetlands that surround the town, including several threatened species. Birdlife International (a worldwide association of bird conservation organisations) has designated the Wakkerstroom area as an Important Bird and Biodiversity Area, meaning that it has significance above that of surrounding areas, as defined by specific criteria. Many houses in town have been bought and renovated by birding folk, often to be rented out as accommodation for visiting birders. There are at least two birding shops in town, plus a Birding South Africa place just out of it, which also has a shop, as well as its primary purpose of running a training school, whose graduates now work as birding guides in Wakkerstroom.

Tall tails and true

Widowbirds form a pretty dramatic group within the large African and southern Asian Family of weavers, named for their superb woven grass nests, usually domes with side openings, often built in colonies. In most species, the male is the weaver, and his skill in the art is judged by the females: only if his nest meets her exacting requirements will he get to mate with her. Male Long-tailed Widowbirds take a quite different approach, however. He holds a territory, within which he perfunctorily twists a few blades of grass together, all that is left of his ancestors' nest-weaving prowess. Instead of providing a nest, he competes with the neighbouring males for the females' attention by means of his amazing display flights, though he will also display from a perch, where his tail is equally visible (Craig 2017). One might think that a skilfully constructed nest might be of more value to a female widowbird than a spectacular tail, but apparently she doesn't see it that way. Having fallen for 'the male with the tail', she builds her own nest in his territory, along with up to four others similarly infatuated.

Thanks to some clever experiments (albeit a bit tough on some of the males!) in the early 1980s, we know that females make their choice as to whose territory to nest in simply on the length of his tail! In that experiment, territory-holding males were caught, and some had a length of their tail removed and taped onto that of another male. The birds were tested in groups of four – one with a short tail, one with an even more extended one, and two with the original length tail, though one of those had his tail snipped and then reattached, just in case that process had an effect. The test was to compare how many nests were added to each territory after the procedure. Unsurprisingly, the short-tailed males did worst, but less expected was the finding that the males with ridiculously long tails were a great hit with the females, even though they'd never encountered such a phenomenon before! (Andersson 1982). (You'll be pleased to know that at the end of breeding season he moults into a drabber short-tailed version of his breeding glory, so that next season the tail-cropped males were back to their normal gorgeous and popular selves.)

There has been intense debate on the subject of 'sexual selection' since Charles Darwin pondered it, observing that some males carry such extreme ornamentation that it must be detrimental to their very survival. It wasn't until Malte Andersson's experiments on the widowbirds, however, that someone actually sought experimental evidence that females did in fact prefer such over-the-top expressions of machismo. In the case of the widowbirds, the large tail could have been to intimidate rival males, but though he will chase angrily an intruder, the tail isn't spread during that pursuit.

It seems that in showing off his ridiculously exaggerated nether appendage he is saying something along the lines of 'see how I can survive despite carrying this burden around behind me – there's your proof that I have great survival genes that would be perfect for your chicks'.

We can hardly discuss interesting tails and sexual selection without mentioning peacocks – technically either the Indian (Blue) Peafowl or the Green Peafowl of South-East Asia, but we can afford to follow familiar usage here and continue to refer generally to peacocks. (There is also the little-known Congo Peafowl, but it doesn't have a particularly impressive tail and the following material refers to the Asian species.) The peacock's 'tail' is actually not really a tail at all, but comprises enormously extended tail coverts – the usually short feathers that cover the base of a bird's tail above and below – for which the term 'train' is generally used. In a twist on the normal state of affairs, the actual tail feathers merely support a peacock's coverts. Although Malte Andersson and colleagues pioneered the experimental study of sexual selection with his Long-tailed Widowbird work, others have carried it on with studies on peacocks (mostly on captive or feral birds).

Marion Petrie and colleagues studied a feral population in southern England. They found that females inspected a series of males, and invariably mated with the one with the most eye-spots in the train (Petrie *et al.* 1991). Just to make sure it was the eye-spots that the females were basing their judgement on (as opposed to some other factor correlated with elaborate trains), they snipped some spots from the train of successful males during winter, and next spring they were significantly less successful (Petrie and Halliday 1994). The researchers were also able to do what Andersson could not with wild birds: they caged eight males whose mating success over the past two seasons was known, and randomly allocated four females to each. Even the females that drew the short straw in males mated rather than produce no chicks at all, but they weren't going to give their all for a bloke they knew was below par – after all, they only had to check out his tail. Those finding themselves with proven successful males laid more eggs than their less fortunate sisters (Petrie and Williams 1993).

But were all these picky females right in their judgement? It would seem that they were. One winter, two foxes managed to penetrate the perimeter fence and kill five peacock males; while only a third of the males had failed to achieve any matings the previous year, four of the five deaths were in this group. The numbers are small, but statistically significant, and highly suggestive (Petrie 1992). Those that fail to impress females are either less fit than others, or lack some survival skills – either way, they are not the best choice as father of your chicks, and females know that. They are right about that too: the eggs from the experiment above were taken and hatched under identical conditions in an incubator, and the chicks released into the wild situation at an appropriate time. Offspring of the successful males (the ones the peahens preferred) survived better than those of the also-ran males (Petrie 1994).

A longer tail is an impediment in the wild because it is more liable to be seized by a predator, and is heavier to drag around. As with the widowbirds, the male peacock is proclaiming his superior fitness and survival skills with his ability to overcome the handicap.

Is he telling the truth though in so proclaiming his superiority? Again, experimental evidence comes down on his side. A French team led by Adeline Loyau went straight to the internal health of peacocks by checking heterophil levels in the blood (these are cells which are produced in response to infections). The study found that peacocks with the most eyespots in the tail and the most displays per hour had the lowest heterophil levels. Conversely, if they injected a successful bird with lipopolysaccharides, or endotoxins, which in the blood produce a strong immune response called endotoxemia, his display rate fell. *But,* males with more eyespots were better able to resist such attacks and to continue to display at a relatively high rate (Loyau *et al.* 2005)!

(I should acknowledge that, more recently, a Japanese study has sought to dismiss all previous studies, including the ones cited above, because their own study, reported in a provocatively titled paper that directly challenges the title of Petrie *et al.* (1991), didn't show such results (Takahashi *et al.* 2008). However, as some of the previous authors have noted in a reply (Loyau *et al.* 2008), all this shows is that the birds in that particular study seem to have responded differently for reasons unknown, not that all the previous studies must be mistaken. Scientific disagreement is rarely dull!)

Another excellent example of a tail hugely modified into a display tool is that of the Superb Lyrebird: a large, very primitive songbird of south-eastern Australian wet forests. (The closely related Albert's Lyrebird of the temperate rainforests of the New South Wales–Queensland border ranges has a similar, but less extreme, rear adornment.) The male's 16 tail feathers are like those of no other bird's tail. The outer two are long and broad, curving outward into S-shapes with broad black tips and semi-transparent chestnut bars along their length. Twelve inner feathers are filmy and lack barbs, so they are loose and open, grey above but silvery below. Two central ones are long and stiffly quill-like. Normally they are all carried in a long sheath-shape behind him, but, when he is displaying, the whole structure is thrown forwards upside down over his head, displaying the shimmering silver underside. At no point is the tail carried so that the big outer two feathers form a lyre-shape (which gives the birds

The evolution of the tail

Archaeopteryx and other early dinobirds unsurprisingly had long, floppy feathered reptilian tails with up to 23 articulated bones. These were good for steering when running across the ground, but not so good in the air. Moreover, they would have been heavy. It is then also unsurprising that, in a relatively short period of time, the tail shortened dramatically, both by a reduction in the number of bones, and by the shortening of individual bones. A chook, for instance, has just five short tail bones and a pygostyle: a short rod comprised of a few bones fused together. Along with associated fat and muscle, it forms the uropygium, which on the (invaluable, for anatomy) roast chook appears as the 'parson's nose'. This supports a stiff fan-shaped tail of strong feathers (the remiges) all arising from the same point, with huge aerodynamic advantages over the floppy ancestral tail. The oldest known 'modern' bird tail belonged to *Hongshanornis longicresta*: a small bird from 125 million year old Chinese deposits (Chiappe *et al.* 2014). The tail has a key function in efficient flight, but, as long as this is not compromised, it can be used for other purposes or, perhaps more accurately, some compromise is permissible, as with the widowbirds and peacocks, if the purpose is sufficiently important. Lyrebirds can scarcely fly at all, being content to glide downhill and walk up again afterwards.

their name): that was the artefact of a British taxidermist, which been perpetuated by artists.

Some nightjars are also given to extreme caudal exhibitionism. The extraordinary Lyre-tailed Nightjar of the tropical Andes has a pair of outer tail feathers up to 80 cm long. Groups of males gather to flaunt their tails to prospective mates by flicking them provocatively while hovering above the ground (Cleere and Kirwan 2017).

Tails have been put to other purposes too. Groups as disparate as woodpeckers, Northern Hemisphere creepers, South American woodcreepers and some swifts have evolved stiff pointed shafts that are used as props while clinging to a vertical surface – extracting prey from bark crevices for most, or clinging to a cliff while nesting in the case of the spine-tailed swifts. Here, the vane (the 'feathery bit') simply ends before the tip of the shaft.

Australian Brushturkeys (big forest-dwelling mound-dwellers of eastern Australia) have their tail feathers arranged, or least held, to form a vertical fan, presumably for steering when running through dense vegetation. Willie Wagtails (a common and familiar member of the fantail Family throughout the whole of mainland Australia) hunt mostly on the ground and constantly flick their tail from side to side to scare up insects from the grass.

Tails, like tales, can even be heard: some species of snipe have a pair of stiff outstretched tail feathers that vibrate as they dive to earth, making a sound often described as 'bleating'.

Barkly Tableland, Northern Territory: grassy expanses

We had driven for ages, seemingly, through a loose endless flock of Australian Pratincoles, springing into the air from the roadside and bouncing lightly over the blond grasslands with their loose long-winged flight, broad chestnut breast band and red bill, legs trailing beyond the tail-tip.

Given the plight of southern Australian native grasslands, reduced to fragments and rarely protected, I find it most exhilarating to drive the vast grassy distances of the tropical Barkly in the north-east of the Northern Territory. The tableland comprises over 100 000 km^2 of land so flat that there is less than 50 m of altitude variation over the entire area. The Barkly Highway, which runs east from Tennant Creek for over 400 km to the Queensland border, only enters the true Barkly Tableland in the far east of its route. To see the great grasslands properly it is best to drive north from the Barkly Roadhouse on the Tablelands Highway. (Look for Tennant Creek on Google Earth, then go east – the clay soils of the tableland show up clearly against the red sands to the south and west.)

The cracking clays expand and contract with heat and rain, opening deep crevices and closing them again, so that the roots of tree saplings are ripped apart; only the tough tussocks of Mitchell Grasses can establish there.

I have another motive for driving the Barkly whenever I get a chance to: it is a stronghold for the Letter-winged Kite, a rare bird of prey that has hitherto avoided me. It's strange having a *bête noir* that is white! (I've alluded to the intriguing breeding strategy of this bird previously – see page 126.) On this occasion, a hovering white raptor caught our attention and I pulled over rapidly, making sure we were well away from the rushing road train behemoths that ply the highway. Not for the first time, however, the lovely and oddly gull-like hawk turned out to be a Black-shouldered Kite, a common and familiar bird throughout mainland Australia. It and the Letter-winged are two of four similar, and closely related, species of the genus *Elanus* found throughout much of the world, though only in Australia is there more than one.

Wind hovering

Never mind, I never really expect to see Letter-wings anyway (which is maybe why I don't!) and a hovering hawk is always worth admiring. The *Elanus* kites tend to stand almost erect in the air, compared with the other great wind-hoverers, the kestrels, a subgroup of a dozen or so small falcons. We looked earlier at 'true' hovering (see page 51), the doyens of which, the hummingbirds, hang in the air by remarkable and hugely energy-expensive wing movements, rapidly pushing back and then forwards again with equal force, as well of course as resisting gravity. The *Elanus* kites and kestrels don't do this – they're probably too big – but they do something equally impressive in its own way, in balancing themselves against the wind, using constantly adjusted wind beats and fanned tail to neutralise its impact. 'Windhover' is an old English name for the Common Kestrel (see Photo 30).

Moreover, because they are hunting – wind hovering is a way of perching wherever you think there might be lunch below, irrespective of whether there's a tree there – hovering birds need to have a clearly focussed view of the ground, for which they need a still head. Accordingly, they go to great trouble to fix the position of their head in the air, while the body moves as required, much as a bird watching the ground from a swaying branch does.

Legs dangling down, the Barkly kite suddenly dropped towards the ground, then resumed hovering now only 20 or so metres above the ground. Suddenly it parachuted down, wings erect, scuffled on the ground and flew up again with a rodent dangling from its claws; out here it could well be the Letter-wing's favoured meal of Long-haired

Rat. With no trees around, it flew to a roadside post, where we could see through the binoculars how it was still rhythmically clenching and loosening its claws, crushing the chest of the little victim before dismembering it.

Far away at the coast, terns have similarly shaped long pointed wings, and are also hovering experts, balancing against the wind above the waves before plunging down onto fish or shrimp. Unusually in its Family, the Pied Kingfisher of Africa and Asia also hovers above the water to hunt: in still conditions, it can even 'truly' hover for short periods of time, without relying on the wind.

Soaring birds ride the wind; wind hoverers skilfully lean on it.

Other memories of woodlands and grasslands

Like grains of pollen shaken free from a grass head by a passing breeze to drift over the plains and drop arbitrarily to earth, often to perish, sometimes to land on a receptive flower to start new life, here are some images from grasslands and woodlands:

- Driving in pre-dawn light through the bare understorey of dry broad-leaved woodland in Benoué National Park in central Cameroon, the group's primary aim is the poorly known Adamawa Turtle Dove, which may come to drink at a pool on the Benoué River at sun-up, so the pause for one of the highlights of the trip felt all too brief. Perched low on a sloping trunk were two dimly seen huge black birds, more than a metre high, with enormous bills topped with strange knobs, long legs and blue faces. The male had a red throat pouch, the female a blue one. It would be another 2 years before I had a chance to properly enjoy Abyssinian Ground Hornbills – and the dove didn't show.
- The channel country of south-west Queensland is a remarkable place: vast expanses of cracking black soil plains of Mitchell Grass braided with hundreds of stream channels, sometimes dry with occasional deep permanent holes, at other times overflowing to form an endless wetland. On a broad grassy stage between two channel crossings, two lines of big grey cranes faced each other in

a wild and wonderful group dance, a member of each pair on either side. They tossed clumps of grass into the air and tried to catch or stab them as they fell, leaping into the air and touching lightly down, stretching high and bugling: a magnificent blast of sound echoing across the plains, enabled by a long coiled trachea that part-fills the chest cavity. Dancing Brolgas are one of the joys of life.

- An afternoon drive in the short-grassed savannah of Murchison Falls National Park in western Uganda, scattered with big palms and hardy acacias, is an experience to open eyelids drooping after many long days on the road and many early mornings. Groups of African Bush Elephants appear in the landscape, Giraffes process across the stage, Uganda Kob, Oribi, purple-rusty Hartebeest, Waterbuck, Bushbuck, Warthogs and rusty Patas Monkeys make their appearances. And, at last, 2 years after my frustratingly fleeting pre-dawn experience of them, three magnificent Abyssinian Ground Hornbills stalk slowly across the plain – and are almost upstaged! Behind them is another big bird, solitary this time, looking even more unlikely than the hornbills. A Secretarybird is a highly specialised bird of prey, warranting its own Family – and no wonder! It is a long-legged, long-necked, long-tailed grey bird with a disarray of long black feathers sticking back from its eagle-like head, black wings, tail and half-trousers. It mince-shuffles along watching for prey, especially snakes, like a tall person trying to maintain dignity while wearing very loose shoes. I am totally entranced.

References

ACT Government (2004) *Woodlands for Wildlife: ACT Lowland Woodland Conservation Strategy. Action Plan number 27.* Environment ACT, Canberra.

Allentoft ME, Rawlence NJ (2012) Moa's Ark or volant ghosts of Gondwana? Insights from nineteen years of ancient DNA research on the extinct moa (Aves: Dinornithiformes) of New Zealand. *Annals of Anatomy – Anatomischer Anzeiger* **194**(1), 36–51. doi:10.1016/j.aanat.2011.04.002

Andersson M (1982) Female choice selects for extreme tail length in a widowbird. *Nature* **299**, 818–820. doi:10.1038/299818a0

ANU (2016) 'Critically Endangered honeyeater population discovered'. 18 November 2016. Australian National University, Canberra, <http://www.anu.edu.au/news/all-news/critically-endangered-honeyeater-population-discovered>.

Australian Town and Country Journal (1896) Australian birds. Orchard and vineyard pests. No. X. *Meliphaga phrygia*, Latham. *Australian Town and Country Journal* Saturday 2 May 1896, p. 22, <http://trove.nla.gov.au/newspaper/article/71245159>.

Baker AJ, Haddrath O, McPherson JD, Cloutier A (2014) Genomic support for a moa-tinamou clade and adaptive morphological convergence in flightless ratites. *Molecular Biology and Evolution* **31**(7), 1686–1696. doi:10.1093/molbev/msu153

Chiappe LM, Bo Z, O'Connor JK, Chunling G, Xuri W, Habib M, *et al.* (2014) A new specimen of the Early Cretaceous bird *Hongshanornis longicresta*: insights into the aerodynamics and diet of a basal ornithuromorph. *PeerJ* **2**, e234, <https://www.ncbi.nlm.nih.gov/pmc/articles/PMC3898307/>.

Cleere N, Kirwan GM (2017) Lyre-tailed Nightjar (*Uropsalis lyra*). In *Handbook of the Birds of the World Alive*. (Eds J del Hoyo, A Elliott, J Sargatal, DA Christie and E de Juana). Lynx Edicions, Barcelona, Spain, <http://www.hbw.com/node/55247>.

Craig A (2017) Long-tailed Widowbird (*Euplectes progne*). In *Handbook of the Birds of the World Alive*. (Eds J del Hoyo, A Elliott, J Sargatal, DA Christie and E de Juana). Lynx Edicions, Barcelona, Spain, <http://www.hbw.com/node/60997>.

Department of the Environment (2017a) *Dromaius baudinianus* in Species Profile and Threats Database. Department of the Environment, Canberra, <http://www.environment.gov.au/cgi-bin/sprat/public/publicspecies.pl?taxon_id=26183>.

Department of the Environment (2017b) *Dromaius ater* in Species Profile and Threats Database. Department of the Environment, Canberra, <http://www.environment.gov.au/cgi-bin/sprat/public/publicspecies.pl?taxon_id=66724>.

Dolman G, Joseph L (2015) Evolutionary history of birds across southern Australia: structure, history and taxonomic implications of mitochondrial DNA diversity in an ecologically diverse suite of species. *Emu* **115**, 35–48.

Fuller E (2002) Foreword on extinct birds. In *Handbook of the Birds of the World*. Vol. 7. (Eds J del Hoyo, A Elliott and J Sagatal). Lynx Edicions, Barcelona, Spain, <http://www.lynxeds.com/hbw/foreword/hbw-7-foreword-extinct-birds-errol-fuller>.

Garnett S, Szabo J, Dutson G (2011) *The Action Plan for Australian Birds 2010.* CSIRO Publishing, Melbourne.

Gould J (1865) *Handbook to the Birds of Australia.* Vol. 1. John Gould, London, UK.

Harshman J, Braund EL, Braun MJ, Huddleston CJ, Bowie RCK, Chojnowski JL, *et al.* (2008) Phylogenomic evidence for multiple losses of flight in ratite birds. *Proceedings of the National Academy of Sciences of the United States of America* **105**(36), 13462–13467. doi:10.1073/pnas.0803242105

Heupink TM, Huynen L, Lambert DM (2011) Ancient DNA suggests Dwarf and 'Giant' Emu are conspecific. *PLoS One* **6**(4), e18728, <http://journals.plos.org/plosone/article?id=10.1371/journal.pone.0018728>.

Higgins PJ, Peter JM, Cowling SJ (2006) *Handbook of Australian, New Zealand and Antarctic Birds.* Vol. 7(A). Oxford University Press, Melbourne.

Lederer RJ (1972) The role of avian rictal bristles. *The Wilson Bulletin* **84**(2), 193–197.

Lindenmayer DR, Margules CR, Botkin BB (2000) Indicators of biodiversity for ecologically sustainable forest management. *Conservation Biology* **14**(4), 941–950. doi:10.1046/j.1523-1739.2000.98533.x

Loyau A, Saint Jalme M, Cagniant C (2005) Multiple sexual advertisements honestly reflect health status in peacocks (*Pavo cristatus*). *Behavioral Ecology and Sociobiology* **58**(6), 552–557. doi:10.1007/s00265-005-0958-y

Loyau A, Petrie M, Saint Jalme M, Sorci G (2008) Do peahens not prefer peacocks with more elaborate trains? *Animal Behaviour* **76**(5), e5–e9. doi:10.1016/j.anbehav.2008.07.021

Mitchell KJ, Llamas B, Soubrier J, Rawlence NJ, Worthy TH, Wood J, *et al.* (2014) Ancient DNA reveals elephant birds and kiwi are sister taxa and clarifies ratite bird evolution. *Science* **344**(6186), 898–900. doi:10.1126/science.1251981

Parker SA (1984) The extinct Kangaroo Island Emu, a hitherto unrecognised species. *Bulletin of the British Ornithological Club* **104**(1), 19–22.

Petrie M (1992) Peacocks with low mating success are more likely to suffer predation. *Animal Behaviour* **44**, 585–586.

Petrie M (1994) Improved growth and survival of offspring of peacocks with more elaborate trains. *Nature* **371**(6498), 598–599. doi:10.1038/371598a0

Petrie M, Halliday T (1994) Experimental and natural changes in the peacock's (*Pavo cristatus*) train can affect mating success. *Behavioral Ecology and Sociobiology* **35**(3), 213–217. doi:10.1007/BF00167962

Petrie M, Williams A (1993) Peahens lay more eggs for peacocks with larger trains. *Proceedings of the Royal Society B: Biological Sciences* **251**(1331), 127–131. doi:10.1098/rspb.1993.0018

Petrie M, Halliday T, Sanders C (1991) Peahens prefer peacocks with elaborate trains. *Animal Behaviour* **41**(2), 323–331. doi:10.1016/S0003-3472(05)80484-1

Phillips JP, Gibb GC, Crimp EA, Penny D (2010) Tinamous and moa flock together: mitochondrial genome sequence analysis reveals independent losses of flight among ratites. *Systematic Biology* **59**(1), 90–107. doi:10.1093/sysbio/syp079

Szabo S (1993) Huia, the Sacred Bird. *New Zealand Geographic* **20**, <https://www.nzgeo.com/stories/huia-the-sacred-bird/>.

Takahashi M, Arita H, Hiraiwa-Hasegawa M, Hasegawa T (2008) Peahens do not prefer peacocks with more elaborate trains. *Animal Behaviour* **75**(4), 1209–1219. doi:10.1016/j.anbehav.2007.10.004

Thiele KR, Prober SM (2000) Reserve concepts and conceptual reserves: options for the protection of fragmented ecosystems. In *Temperate Eucalypt Woodlands in Australia: Biology, Conservation, Management and Restoration.* (Eds RJ Hobbs and CJ Yates) pp. 351–358. Surrey Beatty & Sons, Sydney.

Vágási CI, Pap PL, Vincze O, Benko Z, Marton A, Barta Z (2012) Haste makes waste but condition matters: molt rate–feather quality trade-off in a sedentary songbird. *PLoS One* **7**, e40651. doi:10.1371/journal.pone.0040651

West-Sooby J (2013) *Discovery and Empire: The French in the South Seas.* Adelaide University Press, Adelaide.

White HL (1909) Warty-faced Honeyeaters and Friar-birds. *Emu* **9**(2), 93–94. doi:10.1071/MU909092f

Bird species index

Taxonomy per IOC World Bird List Version 7.2 (retrieved from http://www.worldbirdnames.org/classification/family-index/ on 18 May 2017).

Page numbers in bold indicate photographs.

General index

Page numbers in bold indicate photographs.